Alone
But Never Lonely

BOB FIDLER

 FriesenPress

Suite 300 – 990 Fort St
Victoria, BC, V8V 3K2
Canada

www.friesenpress.com

Copyright © 2016 by Bob Fidler
First Edition — 2016

All rights reserved.

No part of this publication may be reproduced in any form, or by any means, electronic or mechanical, including photocopying, recording, or any information browsing, storage, or retrieval system, without permission in writing from FriesenPress.

ISBN
978-1-4602-8376-9 (Hardcover)
978-1-4602-8377-6 (Paperback)
978-1-4602-8378-3 (eBook)

1. BIOGRAPHY & AUTOBIOGRAPHY, ADVENTURERS & EXPLORERS

Distributed to the trade by The Ingram Book Company

Gus D'Aoust—-Barren Ground (Arctic Tundra) Trapper and Adventurer

Gus D'Aoust and I first met at his home cabin and trading post at the east end of Great Slave Lake near Fort Reliance, Northwest Territories, in 1972. I was on the lake seeking an access route for canoeing through the tundra lakes and the Coppermine River to the Eskimo town of Coppermine on the Coronation Gulf of the Arctic Ocean in 1973. While giving our '72 itinerary to the RCMP in Yellowknife, it was suggested strongly that we talk to "this old trapper, Gus" if we should make it to Fort Reliance, because he would be an invaluable source of information. Already intending to explore nearby Pikes Peak Portage, we headed that way.

When we landed the canoes at Gus's place, dogs barked, and this lean, elderly man came hobbling down the trail to meet us. I think we became lifetime friends with the first handshake, even though neither of us had ever seen or heard of the other. I told him what we wanted, he invited us in, and we didn't leave for more than a week.

Gus liked to entertain visitors, something which he got little chance to do so because of his remoteness. Happy to be one of them, I quickly became very intrigued as his story began to unfold. In short order, I knew

I couldn't get enough of it, and so got my cassette recorder from my pack. Eventually, I filled all the tapes I had with me with his stories, in his words. After that, I took notes on most of the rest of our conversations.

I didn't know at the time just what I would do with this information; but for some reason, I wanted more of his story. We worked out a system to continue our talks in a unique way after I returned to the States. I would send him some blank tapes, and he would arrange to borrow a recorder from an acquaintance at Fort Reliance and record all he could remember about his good old days. Even though I expected very little from it, I sent along the tapes. And, two or three months later, he returned them, loaded with Gus D'Aoust stories.

In 1973, we returned to the Territories to see through our long-planned canoe trip. We put the 180 or so miles behind us from Yellowknife to Fort Reliance and then, of course, stopped in to see Gus. He insisted that we wouldn't survive the rest of the trip and tried to discourage our undertaking it, but promised to see us in Yellowknife should we somehow live to board our future plane charter. I gave him some more blank tapes, told him we had to make it so I could get them back, and took off. We made it—of course—and, true to his word, Gus met us on the tarmac at the Yellowknife airport when we landed. He gave me a hug, said he just knew we wouldn't make it, and handed me more tapes of his life.

In 1974, we returned to Yellowknife as the chartered flight leaving the Arctic Coast the year before hadn't been able to carry our canoes back with us as they had promised. We had left some money with the Eskimos to cover the cost of putting them on a jet to fly them to Yellowknife, which they had done. A friend had picked them up at the airport and had them waiting at her house for us. We retrieved them, made the short trip to Great Slave Lake, and spent the summer there fishing and camping. I had just gotten married and this was a honeymoon in Utopia. Still, there was some disappointment to that visit. Gus was in his native Manitoba when we were there, visiting a sister. We made phone contact with him from Yellowknife through our local friend before departing, however, and managed to see him in Manitoba on our way back to the States. Though we exchanged a few more tapes, this was the last time I would ever see Gus.

The urge to put his life and experiences into print became a reality for me in late 1973 after I returned home from my Arctic canoe trip. Putting things into a somewhat chronological order from the notes and tapes was anything but fun, but I got there. The most trying and difficult part of the task was sorting the names and places he mentioned, and assuring myself of their correct spelling (often difficult because pronunciation was sometimes doubtful). I had tried to pin that down on our last visit by asking him to spell some of the more dubious words, but he would always say, "You spell it" to me. Then when I did, his patented answer was always the same: "That's right. You sure can spell." And so, I can offer no assurances of the spelling accuracy of some of the names and places mentioned. But they were correct in Gus's eyes, and that's good enough for me.

Gus passed away in 1990 at the age of 93. The thoughts and feelings he expressed to me were his alone and have forever held me in awe of the man and his way of life. I have tried to convey these thoughts and feelings as accurately as possible. He felt very strongly about his basic life philosophy, and was quite adamant in the depth it took to express it.

I feel blessed and honored to have walked in many of the same footsteps and paddled some of the same waters he did, while traveling hundreds of miles across the Barrens. I share many of his feelings and hope I have expressed them in a way that honors him. Times, conditions, events, lifestyles, and public sentiments have greatly changed since Gus's time. Some of the things he said, did, believed in, and felt would not be possible today. Gus D'Aoust was an icon of freedom for his time, and I hope I have done him justice.

This was written in 1973, put in a box, and forgotten about until now. If it holds some interest to enough people, I'd like to share Gus. He and his life were very unique, and neither is likely to be duplicated any time soon.

Bob Fidler

Alone
But Never Lonely

✻

CHAPTER I

When God dispersed the human race over this great earth's varied surface, He set aside particular places for particular people. He blessed certain men with the mental and physical abilities they would require in order to cope with the range of natural elements He'd created before man was even a reality. More than that, man was also equipped with the capability to reach out and, through his ingenuity, lay claim for himself to a section of the earth that he might call home.

Men have been reaching out thus ever since. And over my nearly eighty years in this great land of the North, I've borne witness multiple times to man's most basic impulse to search for things and places that improve his situation. Man has always been a natural adventurer and explorer, ever inching forward into new territories in an effort to learn about the unknown, ever seeking that end-of-the-rainbow gold pot. And though most never find it, many find something more valuable, real, and meaningful; something that all men seek, after all: happiness in a way of life.

Through much of my life, I have felt the urge to travel into a new country to pursue my way of life with a spirit of freedom and happiness always providing the fuel. But many times in this quest for life-fulfilling satisfaction the way became difficult; sometimes, almost impossible. There were countless obstacles to be overcome, and I had to arrive at a range of understandings with both myself and others around me when I came upon them. To achieve my goals, I had to make even my feelings for loved ones secondary.

You see, at a very early age, I knew and understood that my life was to be one of hunting and trapping. This was in my blood and burned deeply into my soul. It was an ambition, a force established since it emerged as a childhood dream. Nothing was going to stop me from reaching out for this way of life. I was enough of an explorer and adventurer to do what I wanted and, thanks to all of my stubborn determination, I made it.

And throughout this lifetime of experience and observation, I've noticed a very basic theme. As sure as there are places on earth for certain men, so are there men for certain places. I'm sure this has been stated many times, but I'm equally sure that very few men identify and settle in their perfect places. So many remain in close proximity of their starting points, never testing the urge to venture. I feel sadness for those who never answered this drive.

Many men who have, however, have found their way to Fort Reliance. They've come to visit and to repeatedly inquire how they might find fulfillment way out at the edge of the tree line. Some have referred to me as a hermit. Others have considered me a nomad, seeking a monastic way of life. A few have suggested that I must have felt little love for my mother, father, wife, and children. Once, I was compared to a rebel; another time, someone hinted that I was nothing but simple-minded Barrens trapper.

None of these is even remotely true. Above all else, I consider myself to be a man, neither better nor worse than any other. I sought a way of life I loved and believed in, and found it. All these years I've kept abreast of the way of life and society on the outside. And let no man deny that I've settled in one of the most beautiful places on God's earth. I have a profound faith in Christianity, and the breadth of love taught therein. I've traveled and read and maintained as much as any practical man of the North has.

Through my brothers and I, the name D'Aoust—French, and pronounced "Dow," rhyming with "now"—has become known to the Northwest Territories. There are lots of D'Aousts in Eastern Canada. It's a Duke's name without the money.

This land I have chosen for my home is a vast one indeed. Few people, slightly fewer than 40,000, can truthfully lay claim to being a resident of the Northwest Territories, a land mass that comprises nearly one-third of the total area of Canada. It's big, all right, and still a land for the man with a true pioneer spirit. I know a great many of its people, and they are in a class apart from any other.

In general, the people are crisp of mind and stout of heart. They demonstrate the integrity and honesty it takes to survive in the North. People up here depend upon each other. I can look back over fifty years of this land and its people, and am honored to note that both are still new and shiny. This newness is unique and probably quite difficult to understand in the crowded world to the south.

In recalling the events that go into the making of a lifetime, it would be so easy to exaggerate. But I don't and I won't. When I say something, it's true. By the Jesus, I don't have to lie. I've had so many things happen that sound unreal that I have to be careful when I tell the truth. So, no, I will tell no lies. This country and its people speak for themselves, and I'd never do anything to tarnish their image.

Those of us living lives of remoteness don't talk with boastfulness of our adventures, but with a reverence that lets the undertone of excitement speak for itself. Beneath our brows, toughened by countless blizzards met head-on in the Barrens, lies a humbleness that's real.

Why would the public want to know anything about an old Barrens relic like me? I'm only a trapper living a simple life. I've long been asked when I'm going to write a book about my experiences, and I always say I'm going to start "next winter." But I never do. Besides, my world and life haven't been outstanding—just different.

The biggest reason I came north to the Territories was to be my own man doing my chosen work: trapping. That was always at the top of my mind, even as a small boy. The feeling that just maybe I'd trap and walk where no one else had before pushed me onward.

A few years ago, a group of us old trappers and travelers were invited by the Queen of England to gather at Fort Smith. She shook my hand, too. So we old-timers of course started talking about the old days. Somebody asked if anyone had ever traveled into an area around the Barrens where he hadn't seen a log or tree with an ax mark nearby. We all thought and thought and finally all agreed that, no, we hadn't. No matter where we'd been, someone had always been there before us.

But although I never found absolute solitude, I came closer than most men. The occasional mark of an ax as the only reminder of other men having stopped by before me was acceptable. It proved others had survived and thus provided assurance that I could, too. This vast country of long, hard winters and short, beautiful summers satisfied the hunger I always felt and kept my enthusiasm for reaching out at a high level.

Never have I regretted even for a minute my decision to come to the North. Up here, I've had my freedom. I've always been able to pursue the way of life I most wanted and have been happy while making a living. The North has been the answer to all the dreams and prayers of my early life.

Other youngsters idolized cowboys and generals. My heroes were the parka-clad men of the North who followed their dog teams along the trap lines in search of fur. Trapping took control of my being. It was in the very marrow of my bones. And the country never did anything to change my views about the way of life I'd chosen for myself. I'd come here again if I had my life to live over. But next time, I'd arrive at a much earlier age.

To some men, faith is a shadow cast in many directions. When a man has too many demands and pressures weighing on his shoulders, this faith goes out in streams, sometimes to the point that there's nothing left to which he might return in the event of crisis.

I live away from civilization and a formalized place of worship, but must a man be in a house to gain insight into his belief as he prepares himself for his creator's calling? I've found my belief in God as an entity has been made firmer by my life of solitude, particularly while living in an area of such great extremes as are found beyond the tree line upon the tundra. Out there, the natural elements are never still. They reach out to draw a man into their grasps of death. I always knew to whom my fate belonged, and my faith in God could never have been mightier.

I'm a Roman Catholic. I was born and raised that way and, like most people, stayed pretty close to the religious teachings of my childhood. They're the ones I believe in and know best. I don't mean to imply that I'm overly religious, but I believe in the hereafter. There's a life after this one; I know it.

Incidentally, out there was where I believed the most. That's where I saw and felt things the deepest. It was away from everything, well beyond civilization where so many of the works of the devil hadn't reached. All the robbing and killing that take place where packs of people live sometimes cloud things up. Out in the Barrens, I've seen things and had a lot of time to think. Often, I've prayed and it's pulled me through. Several times, that's all that kept me going.

Many times, I've asked God for help, too. Like in a heavy ground drift or a driving blizzard when I was pushing hard to reach a camp that was impossible to see. All at once, I'd shield my eyes with my hand and there it was. Grateful? Words can't describe how much and I always knew where to send my thanks. So I told Him. And I wasn't ashamed to say it out loud, either.

It's easy to find oneself in a tough spot out in the Barrens. And these spots could cost you your life. This is especially true when carelessness sneaks up on you. Most certainly it helped me to ask for direction. It worked, too.

As a youngster, my idle hours were filled with those dreams reserved for boyhood. Entwined among the countless thoughts and immature plans of this age were highlights—ideas to which my mind returned frequently. Most thoughts were scrapped because mental maturity and undisturbed concentration made than impossible. Still, these were the images in my youth that went into the plotting of my future. Countless times my mind in its flight to boyhood fantasy foresaw adventure, conquest, and riches. Of course, few of these desires were ever realized and most of the places—if they ever existed in the first place—were never visited.

But there were those times when my thinking repeatedly returned to that one glorious subject that highlighted all my dreams. It was a burning, overwhelming desire and I'd have gone to all ends to accomplish it. Becoming a trapper was a bull-headed obsession I couldn't shake.

At no time in my school years was I crazy about composition or English literature. I vividly recall sitting at my desk with a school book propped upright on top of it. It shielded a trapping book I had spread out in my lap. I danged sure wasn't paying any attention to that lesson book because I was too busy. I just couldn't read or study enough about trappers and trapping.

Not many years after I came into the world (Makinak, Manitoba, September 18, 1896), I felt a stir to strike out for new country. My life as a farmer's son was a far cry from the one I'd have chosen. I simply couldn't force my roots into the land enough to secure a firm hold. They kept growing toward the North until, at a certain point, the possibility of my ever becoming a farmer had come and gone. Still very young and with the narrow-minded zest reserved for that period in life, I pleaded with my father to sell our quarter and eighty acres and head northward for the Peace River country. Even though I was at an irresponsible age, I was convinced the magnetic North was tugging at my shirt.

But of course my father couldn't gather the family around him and pull up stakes at the whim of a youngster. And my Dad, bless his soul, worked the hell out of us on the farm. In those days, there was no other way. Ten hours a day for three meals. It sure didn't take long to figure there had to be a better way of making a living, from both the work and wage aspects.

With the fur prices sky high, I struck out one fall after the hard work was done.

A nearby range of hills called the Riding Mountains beckoned. Here, I had the opportunity to find that something extra, something the farm didn't—and couldn't—provide. This trip to these special mountains started me on the trail I would follow for all of my life. The trail forked sometimes, and I made some wrong turns, but in darned short order, I would always work my way back to where I belonged.

Before reaching age seventeen, I'd found an old, abandoned cabin, cleaned it up, and adopted it as my very own. Here, I was able to spend two or three weeks at a time tending traps and hunting. I learned about the outdoors as I went. By this time, too, I had tired of school. I was preparing for the practical job before me. I was now becoming a trapper.

It was at about this time that fortune smiled. I crossed trails one day and became acquainted with a fellow named Dave Williams. He had come in from the south to homestead, and was running a winter trap line. From him, I learned how to make various trapping sets, and added countless valuable lessons to my already growing knowledge and understanding of the bush and its animals. His teaching was very important, and it helped me tremendously for the rest of my life.

"Inevitable" is the best way to describe my next step. I felt too big for the country and, like a restless bear, needed something bigger to conquer. So it was that, against the advice and best wishes of my father, the first real outing of Gus D'Aoust, trapper, became a reality.

By the time I was twenty, I had managed to put together enough money for a stake of grub and some traps. I decided to head in the direction of Churchill along the coast of the Hudson Bay in northern Manitoba. I never did get quite to Churchill, but made it to Kettle River, where I left the train to push on into the bush in pursuit of my trapping goal. For the first time, I was completely alone. I must admit that a series of fluttering doubts crossed my mind.

If I lived forever, I'd not forget that moment and how very small it made me feel. As the train pulled away, I was filled up with fear. Still, I was determined and had never taken on board the meaning of the word "quit." When I finally found the wherewithal to start to put my snowshoes on, I glanced to the side and spotted a set of snowshoe rabbit tracks with the prints of a big lynx right beside them. It was then that any and all doubts and fears vanished for me. In an instant, I felt ten feet tall and tougher than a grizzly. There was fur here, and I could feel nothing beyond the spirit of the chase.

I had already studied my maps carefully, but I pulled them out once more at this point for reassurance. I had marked a location where a cabin belonging to a survey team was located. Since it was winter and their work had stalled, the place would be abandoned. I headed there with enough confidence for ten men and set up in the cabin. It became the base for my operation, with the lynx track providing all the encouragement I needed.

That first winter was a wonderful and enlightening one. Like so many to follow, I spent it alone, but was never lonesome. No young man ever served his apprenticeship with more enjoyment than I. There was enough peace and serenity to last most men a lifetime. The winter passed rapidly and I made a catch.

Before returning to the farm in the spring, I had accumulated $800 worth of fur. Included in the varied catch were mink, weasel, and lynx. I also had the only fisher I ever caught. At that time, I had no dogs and walked my entire line. I was what was called a "pack-sack trapper."

Even though I made plans and had ambitions to trap in the Churchill area again, I never made it back. I started out in its direction the following year, but it never became more than a wishful thought. You see, a little thing called World War I came upon the scene with an unexpected rush and so the trapping urge—at least temporarily—had to be put to rest. My number was called and, with that, I was inducted into Her Majesty's army. I was overseas for eleven months and have never been more confused in my life.

To be quite frank, I didn't like that damned war, either. I didn't believe in men killing other men. Curiously, though, I never thought about getting killed myself. That fact never dawned on me. While over there, I made up my mind about one thing for sure: When I got home, I was heading north to trap. There would be no monkey business; I was going to get the hell away from all the problems.

Until the war interrupted my life, I had never smoked at all. But over there somewhere I bought my first pipe, and so spawned a habit that I've enjoyed since. As the stereotyped image of a man living the outdoor life goes, I was a pipe smoker from that time on.

Down through the years, and especially over the last few, I've read all these things about the evils of smoking. Different ones tell how pipe smoking is bad for your health. But I just can't believe it's true unless there is something else the matter with you. By golly, I've had uncles and cousins all over the place who were smokers, too. Matter of fact, most of them had a pipe stuffed in their mouth all the time. I never saw some of them without their pipes, and probably wouldn't have known them if I had. And most of them lived into their seventies and eighties, so smoking couldn't have affected them much. They smoked the hard stuff, too—"Canadian leaf."

No, sir, I don't believe all that stuff I've read about smoking. You know, the first thing I smoked was a cigar, and that was when I was twenty-one. By the Jesus, did I get dizzy! That was enough cigar smoking for me. I stuck to my pipe. And all these years, I've smoked, too. I don't fool around. I hit her hard. Just the same, smoking has caused me some discomfort and distress, too. On a few occasions, I've run out of tobacco out in the Barrens. There were no stores within 200 or 300 miles and, believe me, it was a son of a gun.

There were two times every day when I missed tobacco the most. Like everything else, I was a creature of habit. Most mornings, preparations for the day's run were much the same daily routine. Soon after I first got up,

I loaded the toboggan with everything the trip required. Then the dogs were put into the harness, and I was ready to go. Just before hitting the trail, I did something that was a ritual all the years I followed a team. I lit my pipe. Then "march" ("*moish*") and off we'd go. Right then is one time I missed the tobacco. When I was out of smoking material I was restless. Soon after starting along the trail, maybe three or four miles, or if I saw a fur hanging in a trap, I'd be all right. I got better if I had something to take my mind off the pipe.

The other time I missed it was at camp in the evening when all the chores of the day were done. My pipe was like a grand old friend offering enjoyment and company when I directed my attention to it. Those times I had no tobacco, nothing I did or thought about gave me any satisfaction. I was always looking for something. It was that guldanged tobacco.

So, when I returned from the war, smoking habit and all, I exchanged the uniform of the soldier for the clothes of a farm hand. I was needed on the farm and, for a year-and-a-half, I tried to conform to the ways of my father. But try as I did, farming just wasn't the answer. That drive to go north hadn't disappeared during the change from boyhood to manhood. The inner turmoil and discontent were just too much for me. To keep my spirit, I followed my heart.

The decision to strike out for the North had long since been made, but now the time had arrived to see it through. I had a boyhood chum, Gib Haight, and we'd always planned to go together when the time came. So I went to him and shared my plan to leave soon for greener pastures. For some reason, Gib decided to stay behind, and he never made it into the North to trap. Come hell or high water, I was going anyway. So I gave my Dad and family a farewell and headed for the train station, still unsure of my destination.

Two places raced through my mind and I had to make a big decision. Should I go to Kenora, Ontario, or to the Peace River country of northern Alberta? I thought and thought, but couldn't make up my mind. It wasn't possible to go partway and then decide because the two places were in opposite directions. Deciding was proving too much for me. Finally, I stepped up to the platform, took a coin from my pocket and flipped it skyward. The doggone thing struck the planks, spinning, rolling, and then wobbling to a halt. When it settled, Kenora was buried beneath the coin and the Peace River side was smiling upward at me. It's a little frightening, as I look back, to realize that a man's life can be determined by the mere turn of fortune found in the toss of a coin. At times, I've wondered what might have been if the other side had shown itself.

Anyway, the ticket was purchased and the train was boarded. I had not the slightest premonition, but that ride was to lead toward the country that would be home for the rest of my life. That was in 1919 and I was almost twenty-three. I knew, from my first glimpse upon the country just south of Great Slave Lake, that I was getting very close to my peace on earth.

I've had my freedom in the great spaces of the North. I felt it as soon as I got here and that feeling has grown stronger with the passing of time. Before ever coming into the country, I knew I was going to make money—that fact was bloody obvious. I was a trapper, fur was plentiful, and prices were high. Still, money wasn't the whole reason for coming into the country, and it wouldn't have brought me here by itself. Very simply, the pure North beckoned as the one place on earth where I'd find the most happiness and peace of mind.

Having left the train en route to the Peace River, I had to get settled and decide on my next step. I worked for a while at various jobs afforded by the construction of a crossing over the river. I bided my time here, spreading gravel and working on the sluice wagons. But all the while across the way and before my very eyes, a world of excitement and high adventure was unfolding. I watched impatiently and was awed as scores of trappers and homesteaders came down the river in canoes and on rafts of driftwood and logs.

For hours on end I stood with my eyes on the river and these people following their dreams. I saw some sights, let me tell you. Tents, dogs, cook stoves, and everything in a man's life floated past me on those rafts. Some were enclosed with a fence around their top; others were bare and unprotected.

Out into the four-mile-per-hour current they'd head, with enthusiasm and expectation showing the way. Soon, they'd be around a bend and out of sight, leaving me to wonder—over and over—what would become of them. Many of the rafts were built right near the crossing by builders impatient to get the work completed so they might get on with the journey. It was some 540 miles to Fort Fitzgerald, but the gritty travelers didn't bat an eye. Containing myself in their sight was difficult, and I waited with all the patience my mind could muster.

The waiting continued throughout the fall. Then on through the winter, as the cold North froze, closing the water highways that led beyond the railroads. Finally, the days began to lengthen and then to grow warmer. After what seemed to be an eternity, the thaws came.

As soon as the ice went out in the spring, we struck out down the Peace River. Understand, too, that we weren't very far behind the thaw. I remember the big red scow we were riding. It was possibly twenty-five-feet long and a rugged boat if ever there was one. Eleven of us boarded her for the long ride. A huge timber was cut in the general shape of a paddle and acted as the sweep; a kind of rudder arrangement. We took turns at it, and I was the one in control when we shot the Vermillion Chutes. It didn't take me long to yell for help when we started through, either. I remember I wasn't so scared at the time because everything happened too fast for thinking.

Around the deck we'd built a fence to hold loose objects aboard in case we tossed about. A bloody good thing it was there, too! We started through the Chutes in a fog which was a danged stupid thing to do. That place was a devil's hole anyway, and its roar could be heard for twenty miles when the wind was right. To say the very, very least, it was as spooky as the shadows of hell.

After I yelled for help, three men came to assist with the sweep. Every bloody one of us fell down only to have that beautiful fence catch us all. Otherwise we'd have gone overboard and it would've been curtains for sure. We zoomed through there so fast everything was a big blur. By the time we reached the end, everyone and everything was wet and beaten up. Old Captain Myers wasn't talking so very loudly, either. That bugger knew danged well we'd had a close call and was white around the gills. I had to agree with him.

I'd made friends with a man named Joe Michlau, a boat mate with whom I shared a great deal. Our interests were similar and we both rode the river in that scow for the same single reason: we needed a grub stake so that we could go trapping. We had heard work and fair wages could be found at the sixteen-mile portage between Fort Fitzgerald and Fort Smith. That's where we were headed.

I had plenty of time for studying the land as the boat made its slow way toward the place of my dreams. Wild game was plentiful because the plane and automobile hadn't yet brought the masses of people with whom the animals would find it difficult to share the land.

At that time, everything was pretty much the way God made it. There wasn't any of this dirt, filth, or pollution. My eyes never stopped as I tried to soak up every sight over that whole route. And I wondered about a lot of things during that month. Especially when we'd float past cabins that were no closer together than twenty or thirty miles. One question hit me very hard in these early days: How in the world did those people live? There was no doctor anywhere close to them. In fact, there was nothing. And so, believe me, that feeling of being alone hit me at the start. But, you know, later on, I learned to control my thinking and never got lonely. I simply never allowed my mind to run away from me.

One instance that stands out in my recollections of this first trip into the North happened because those of us on the scow had gotten tired of the food and wanted some fresh meat. We had seen game along the river at frequent intervals, and itched to improve our rations. And then one day on a straight stretch of river, we saw a moose well ahead of us, standing on a sand bar. He was a beauty and I could just taste the sweetness of that meat. Asking first if the boat could be stopped to pick up the animal, and having been given assurance that it would, we decided to shoot him.

Joe was elected to do the killing. When we got within range, he dropped the moose with one well-placed shot. We were happy, but the feeling didn't last very long. The man who was handling the sweep refused to stop the boat. He wouldn't steer it to shore. Gosh, I got sore. It turned my stomach to think about wasting that meat—any meat for that matter. When that fellow said he'd stop and then changed his mind, I felt like tearing into him.

After about a month, we arrived at the sixteen-mile portage. Joe and I contracted to cut 200 cords of wood. We did it the hard way and struggled some, but upheld our end of the bargain. We even cut forty cords extra and sold it on the side. By golly, we had the money we needed to outfit ourselves for the following winter's trapping.

I'd have done anything to get that stake. It had been a lot of wood, but I was danged sure going to cut it—flies and bugs be damned. I knew that, when I got a good outfit together, I would be in the height of my glory. Good dogs, equipment, and grub stake: that's what I lived for. And I worked hard all my life to make sure I had the tools I needed to carry on with my trade.

CHAPTER II

I stayed to work and trap the area of the Slave River to the south of Great Slave Lake until 1929. At that time, this was wonderful, uncrowded country. My trapping labors were proving profitable enough to provide a good living. That in itself was all I needed to be able to continue with this job that I never once considered work.

I was always at ease with the world because, out in the wilderness, I thought of myself as a man living within the guidelines of his destiny. This life was the natural way for me to live. I answered to no one and depended solely upon my own abilities and knowledge. I knew I was carving my own niche in the world, and I was having fun doing it.

Once free to chase fur, I shuddered with every thought of a shovel or plow. To that particular way of life I'd never return. I knew my living would be made by my ability to place a trap where some animal might chance to step.

Let me say, too, for those to whom the taking of an animal in this fashion is disgusting: That this was fur country in an era of trapping. It was a tough life of competing with nature and her animals. Besides, if a man was to live in the country, trapping was the only livelihood available. Trappers came into the North before the prospector or homesteader, bringing out with them not only their catches, but words and descriptions about a vast area inhabited primarily by Indians in the south and Eskimos in the north.

I'm not exactly sure where I fit in, but most of the trappers were more than gatherers of fur. They were explorers and adventurers with a lust for the great outdoors. Even though the calendar had slipped around to the twentieth century, the cold, remote North was still enough of an uncertainty to make most men back away before tackling it. Not until the arrival of the float plane—and, with it, the amazing bush pilot—was the far North brought in close contact with the ways of civilization to the south.

The trapper was the first consistent link between the North and the people to the south. I've read of the early explorers and traders who came into the country seeking water passages as shortcuts through North America, and minerals such as copper to line the coffers of industry. But what could these people offer to the world waiting to learn about the mysterious North? Almost nothing. It was the fur trapper—who spent three-quarters of each year getting into, living within, and coming back out of the North—who could truly convey to others the moods and feelings they wanted to hear.

I've been proud to carry on this tradition. I've never thought of myself as some kind of hermit seeking to hide from the civilized world. But I had great aspirations to trap, and I wanted to go where there was fur. I always tried to maintain contact with the outside world. And I've done this as well as any man living away from it could. I didn't let the world escape. But I perhaps have escaped it, to an extent.

Like all men with a strong will and a powerful wish for independence, I was susceptible to few things that might alter my planned course. But I wasn't able to avoid everything. It was possible to pierce even the thick armor of *my* bull-headed determination, and it was done by nothing more than the look from a pair of eyes belonging to a woman. I guess I was no exception to this rule of nature either; I joined the traditional world and took a wife.

Though I have many thoughts about my marriage, there are few words to describe it other than doomed from the beginning. The match-up was a mistake and we both knew it right away. I was a man of the North living an outdoor life and couldn't live in the South where she was happy. And she hated my world in the North just as badly. In the beginning, she tried to accept my way of life, and spent two years with me. She eventually called them wasted years, and maybe they were for her. The life up here isn't for everyone. If it were, there'd obviously be more people.

I also tried her way in the South, but that didn't work, either. I felt helpless, like an empty, useless man. Even though we had two very lovely little girls, I still couldn't adjust.

I'd just completed a three-year job at Wood Buffalo National Park on the border of the Territories and Alberta. This was during the infancy of this vast park, and I saw it expand to its present size. But fascinating as that was, I disliked every moment of my tenure with this operation. This was one of the times I took the wrong fork in the trail.

They gave me seventy-five dollars a month and grub for one for my duties. That wasn't bad money in the twenties, but still, it wasn't worth it. This was especially true since I was a trapper and good marten pelts were worth up to forty dollars apiece. But I was trying to live the life of a family man, and taking a chance that everything would work out for the best.

At the time the government set the land aside for a park, there were only a few hundred big wood bison in existence. They were that close to becoming extinct. The officials brought in some plains buffalo in the hopes that they would mingle with and breed to the big fellows. This was a big project at that time, and no one knew what the results would be.

Several plains buffalo arrived at the releasing point by barge. Before they were unloaded, the barges were turned round and round in the river. The whole idea was to confuse the buffalo to the extent that they wouldn't head back south after being released. Some smart fellow determined that this maneuver would remove their sense of direction. When they figured the buffalo had been spun around enough, the barges were beached and the animals released. A mile-long chute had been built for the buffalo to run through. The experts figured the circling barges and mile-long run in the wrong direction would solve all their problems. But they forgot to tell the buffalo, because when those buggers hit the end of the chute, they made a quick turn and headed straight south. And they were going like hell. They weren't dumb, these guys. Home was to the south and they knew it.

After this experiment failed, the need for my job was created. I was hired to keep an eye on the plains buffalo who showed such a homing instinct. I was out on constant patrol to try and determine just how many had completely left the country to return south to the plains, their natural range.

Those fellows kept on going, as steady as you please, until reaching the Peace River. Here, they stopped and absolutely refused to cross. The government was somewhat taken aback by it all and, not knowing what else to do, decided to make the park bigger. With this move, the park now included the range of *both* the wood buffalo *and* the newly adopted home of the plains buffalo.

One of my tasks was to maintain a census on the herd to determine the mortality rate. After some careful study, it was figured to be seventy-five percent. The bloody wolves were having a picnic like you've never seen on those buffalo, who were getting killed right and left. I know for a fact they slaughtered them because I found kill after kill right close to my trail. It's a wonder any of them survived, but nature took care of things. Today, there's a sizable herd, and a few are hunted each year to keep the population in check.

It was 1929 when I hit the point of knowing I had to adjust to married life. I had determined to make a sincere effort at fulfilling my marital responsibility, and so left the North and headed for Edmonton, Alberta. This was to be a futile attempt at conforming to the life of a working husband and father. A regimented schedule every day had never been my idea of pleasure, and this jaunt did nothing to change that belief. It most definitely wasn't because I was lazy—and anyone who tried to follow me around could attest to that—it was because I was working for another man, and this I hated with a profound and lasting passion. My freedom was gone.

We lived a long way from my place of employment and one day I was late for work. Usually I had a ride, but transportation hadn't been available on this day, so I had walked the whole distance. My boss cautioned me about my tardiness in a manner that left a bad taste in my mouth. I was accustomed to an atmosphere of independence, and it took all my willpower to keep quiet. I didn't apologize or promise it wouldn't happen another time.

Being late for work again was inevitable because of the distance and unreliability of transportation. And so, sure enough, it happened again.

This time, the man scolded me with some rather pointed language and tossed in some idle threats. I kept my thoughts to myself, but at the end of that one-sided conversation, I went out the door and never returned. I didn't even return to collect the money I had coming, because taking money from that kind of person wasn't worth it.

For some reason, instant adventure was never more than an arm's length away for me. And it wasn't any different with the other trappers I knew at the time. We had a yen to explore new country, and so tolerated many things that would have made the average man miserable. We prided ourselves on the idea that we were always careful, and that we took no unnecessary chances. I'm now aware that our lives were hanging by a thread each time we entered the Barrens to live alone during the worst seven to eight months of the year.

On the other hand, I didn't mind taking a chance with my money when I thought a little investment might pay some dividends. One day, I came across a magazine advertisement that looked like a good deal. It seemed an opportunity to put up a few dollars for a short period of time, and then to get it back at a high return. Not surprisingly, the offer had to do with fur. By putting up $250, the ad explained, I'd become a shareholder in a fur farm in Grand Rapids, Minnesota.

At that time, the deal looked good and I had the cash. Without any hesitation, I invested the money. But then a couple of years went by without any word from the fur farm or the get-rich venture. I couldn't figure out what had happened.

Honestly, I don't know why I ever sent the money called for in that ad. Only a danged fool would think he'd get wealthy overnight on a small investment, and I suppose I qualified. Even though $250 doesn't sound like much money now, in those days, it was a sizeable chunk. I wanted to find out what had happened to it.

It was then that a sudden lust for adventure took over and sent me on one of the most unusual journeys of my life.

In 1929, there weren't many different ways to travel. But if you desired speed to cross a great expanse of land, the train was usually the choice. Travel over the rails was comfortable and sure, and many people enjoyed the luxury. I didn't let the fact that money was a limiting factor for me get in the way of my own pursuit of this form of transportation. With only ten dollars in my pocket, I struck out for Minnesota—and by train, too. Still, my ticket wasn't terribly conventional. I traveled hobo-style, and hopped trains.

And so one morning I decided to head for the railyards in Edmonton, where I lived with a family that needed that money. After searching around, I found a train loaded and ready to depart for Winnipeg. That was my route, so I hopped on a car and off we went. Everything went smooth as silk and it was an enjoyable ride.

I left this train in Winnipeg because it was to continue east while I needed a train that was going south. I soon found one, and thought, boy, this is sure easy. But I wasn't much more than settled down when the train switched to a siding and stopped. This one would never take me to the border.

Once more I had to look around for transportation, and find it in a hurry. A search revealed that there was only one possibility. A train with cars loaded with grain separators was in line to go to the border. I crawled under one of the things and waited.

A man was traveling with me and had been since Edmonton. He forced himself under another separator. The brakeman was in plain sight when we'd both squeezed ourselves into position, and surely had to have seen us. We had climbed on bold as you please with no attempt at sneaking.

It wasn't time for the train to pull out. We relaxed under our separators during the delay, and the brakeman came over and stood near our car. We tried every way we knew to tunnel deeper, but in those things, it was impossible. The man had to have seen our legs sticking out. He couldn't miss them. But he didn't show in any way that he'd spotted us, so we breathed easier for a minute—a very short one.

For some reason, I peeked out—and what I saw scared hell out of me. Down the track a ways was a bull—a policeman. He was walking along the cars, peeking into and prodding at every little crack and crevice. And he was headed directly for our car. No doubt, he was looking for men like us who were illegally riding the trains. They were doggoned tough on guys headed for the border.

The brakeman walked a short distance to meet him, and stopping a little ways beyond our car. There we were with our feet sticking out, a railroad employee who knew we were there, and a policeman only a stone's throw away. I was scared.

When I heard the policeman ask the brakeman if he'd seen anyone around, I thought we'd had it. But for some reason, he refused to tell on us. He said he hadn't seen a soul. I could have kissed him right then and there. I was really stunned when the policeman said he'd caught thirty guys on trains in the past week, and that all of them had been sent to work on grain farms. And for thirty days, no less.

That kind of penalty had no appeal for me and by the Jesus was I happy when the law walked away—so happy that I helped unload those guldanged separators all along the route to the border. They'd stop and I'd jump off to give them a hand. I figured one good turn deserved another.

We crossed the border without any problems and eventually arrived in Grand Rapids, where I left the train. I soon found the location of the fur farm and figured I was on my way to getting my money back. But when I got to the farm, I was surprised to discover a silver fox operation. A quick glance about told me things hadn't been going well. It was a second-rate project with little chance of improving. There was no money around for sure. Others like me, I soon learned, had also invested money. We were all completely out of luck.

Of all things, they offered me one fox to help cover expenses. I wondered what the hell I'd do with a live fox in a strange country. I refused their fox. I had come to Grand Rapids for money, but there wasn't any to be had. All that was left was for me to head back home.

A train took me toward Duluth, where my money began to run very low. The ten dollars I had started with wasn't going very far. It might have been enough for one person, but I was paying for my friend's meals, too. And we weren't eating very well, either. By now, I was down to change. The only other items of value I had were my elk teeth, a matched set of beauties I wouldn't have taken anything for under most circumstances.

When a man gets hungry, he'll do things he wouldn't do at any other time. And I was hungry. I studied those teeth knowing they were priceless —but I had to eat. In Duluth, I looked for and found a pawn shop. It took some time to find the courage, but I eventually did. I went in not knowing what to expect.

I showed the man in charge the teeth and he studied them. After a while, he asked what I thought they were worth. I told him my circumstance while determining that five dollars would see me through. I told him five would do. He offered four, saying I could later get them back for five. He had me and knew it. I had to take his offer and like it, pleading with him to hold onto the teeth until I could send the money. I hocked those elk teeth and have never felt worse about doing anything in my life.

I honestly never expected to see them again. But when I eventually got back to Edmonton and had amassed the required funds, I sent the money to the pawn shop. Sure enough, in a few days, the teeth arrived.

In the early days, I had collected elk teeth as both a hobby and a way to earn some extra money. The teeth were highly valued as charms by many jewelry fanciers and as an emblem to those belonging to Elks Clubs. The price of the teeth—only the two big ones from each adult elk were worth much—was determined by the amount of brown stain from the vegetation the animal ate. The browner, the better. Also, the pair had to be a matched set; single teeth were of little value.

I used to collect all the teeth I could get my hands on, and had paid as much as twenty-two dollars for a matched set with deep stains. Each pair of teeth was unique in that no two sets were alike. Still, I never once shot an elk just for its teeth, although this was a common practice with some.

This was just a hobby with me, but I did make a little money at it. Some of the teeth were sent to a jeweler in Deadwood, South Dakota. After I left the elk country in 1920, I kept a few pairs for myself, but I stopped collecting and selling them.

After pawning the elk teeth, I hopped a train that appeared headed in the direction I wanted to go. I got into a kind of ore car—one of those things with sides and ends, but with no top. A gondola. Soon after it was underway, the tracks made a bend. I knew immediately I was going the wrong direction. I sure didn't want to go to a mine or wherever it was going, so I crawled up the side of the car and jumped off.

Right there, I learned the value of looking before leaping. I was anxious to get off and had paid no attention to the landing spot. I landed on a steep bank and tumbled head over heels, all the way to the bottom of a deep ditch. I danged near broke my neck.

Fortunately, my head missed all the rocks and trees and I came out of it with only some minor scrapes and bruises. That was my last ride on an ore car. I walked back to the railyard seeking a train going my way. The one I chose was a blind baggage, and it went like hell. The thing moved too fast for me to be comfortable, but I hung on to her. You can bet I hung on tight, too, because my position was right behind the engine, down low near the tracks.

We'd been going awhile when, all at once, all hell broke loose. The whistles started blowing and screaming loudly. They sounded frantic. I was scared, imagining all sorts of things. But there wasn't any way to see what was going on. All at once, sparks came shooting up all around and I didn't know what in the world was happening. The sting of smoke and the smell of heat were nearly unbearable. Thoughts of burning up or suffocating were pretty prominent in my mind.

The engineer had applied the brakes in an all-out effort to stop the train, and they had locked, sending all the hot stuff right up in my face. Pretty soon, there came a thud and a jolt as the bloody train smacked into a herd of cows that couldn't get off the tracks. One of them became lodged behind the engine right next to me. What a mess! That cow wasn't my favorite choice as a travel companion.

Before the train came to a stop, I figured I'd better get out of my hiding place, and fast. When it did, I wasted no time doing so, either. I don't think I've ever been happier to escape a place. First thing I did was check around to be sure everything was alright. When the engineer came down, I offered to assist in removing the cow from under the train. The animal weighed over a thousand pounds, and we had a time getting it out.

The train crew appreciated the help and showed it when I started to return to my hiding place. I was interrupted and led to the passenger car. They told me to climb aboard, which seemed like a good idea. I looked around for a seat and relaxed for a change.

When the train had first stopped, most of the passengers had gotten off to investigate the source of trouble. There was all kinds of confusion, so no one paid any attention to the fact that the passenger list had increased.

For a long time, the conductor kept giving me the eye. Something must have looked suspicious because, a while later he came back and asked for my ticket. I told him I'd never had one. He was nice enough about it and asked where I was headed. I told him I was going as far as this train would take me. Two others came on illegally at the same time I did and just sat waiting.

The next time the train pulled into a station and stopped, we were told to get off. We waited around for another train and, not finding one right away, the fellow who had traveled with me went into town. After a while, he came back with word that we could get a job driving a team. I made it clear to him I wanted none of that business. We split up here and he went back for a job. I told him to do as he pleased, but that I was going home.

After he left, I prowled around before finally finding a train headed north toward the Canadian border. This one was sure to move soon because the car was loaded with green tomatoes. It wasn't refrigerated and the tomatoes would spoil if left standing. I climbed into the open end, tomatoes and all. No one spotted me and I got away with it. I crossed the border into Canada without incident.

I became acquainted on that trip with a good many railroad workers. Only one man gave me trouble. I met up with him after I returned to Canada. In Saskatoon, no less. I had to seek out a different train there because mine had reached the end of its line. While snooping around for one, I came upon this car with the door slightly ajar. I looked inside and three or four guys were already there. I crawled in with them.

One of the yard men spotted me and it wasn't but a few minutes before he came over and opened the door. He told us to get off, but nobody budged. We didn't pay any attention to him. In fact, I acted like I didn't hear him. Since there were too many of us, he didn't press the issue and eventually went away.

But pretty soon he came back, this time bringing along a cop for good measure. Before opening the door, the cop yelled for us to be ready to come out and to forget any funny business. When the door opened, I was reaching for the one on the other side. He noticed this and the bloody guy pulled his gun saying he'd shoot if I made a move for that door.

It wasn't necessary for him to show his gun and I told him so. He made me mad, and without any hesitation, I growled right back at him. I asked who the hell he thought he was and how many times he'd gone without eating for two days.

For some reason, that threw him and he didn't say anything right away. Perhaps he didn't expect me to say anything and maybe he was stumped. Then he acted a little sore and frisked me. The only thing he found was some change, maybe sixty cents. That was all the money I had left.

Once I felt he was satisfied, I came right out and told him I was leaving. He seemed to be a little in doubt about me, but finally turned away to give the others his attention. I chose that time to get out of there. The other guys were buffaloed and he took them in. I heard later that this policeman had some trouble and shot a man. At the time I faced him, I never thought about being shot. But I danged sure was determined not to be sent to work on a grain farm. No way was I going to have any of that, so I took off.

I didn't run like a fleeing rabbit, but walked at a steady pace out in the open across the siding to another train and climbed aboard. Either he didn't see me or had his hands full, because I wasn't bothered.

This time, I rode toward the back of a long line of empty cars. The train stopped frequently dropping a few cars at first one place, then another. Every time some cars were removed, I moved up. I kept moving forward until I was right behind the engine. The engine crew saw me all the while, but never said a word.

They must have felt sorry for me at lunch time. When they pulled out their grub, I was invited aboard to share their meal. That was the kindest display of human nature I'd seen in a long time. I was so hungry at the time I could've kissed them.

After talking for a while, they asked where I was going. I told them Edmonton and then exactly the part of town. I was then put on the back car of the train. You know, when the train got to Edmonton, they made a special stop for me. Right where I wanted to go.

I made that trip in September of 1929. It covered well over 2,500 miles and took ten days to complete. I did the whole thing on ten dollars plus the four my hocked elk teeth brought. It was an interesting but strange trip; one I'll never forget. I never got my $250 back either.

The following winter was a tough one. It was my first away from the North in several. I had committed myself to a way of life in the South in an attempt at becoming a settled family man. But it was failing. No matter how hard I tried, I couldn't adjust. At one point, we got the urge to move, and headed west to British Columbia. Vancouver became home for a short time, but I just wasn't happy. In fact, I was totally miserable. I tried everything to forget my feelings, but the old mind was in the North and wouldn't leave it.

I kicked my rear more than once for ever thinking I could make a go of it outside the North. I should've known the city was no place for a man like me. But for the sake of my wife and the girls, I had to make an attempt. As I look back, it was a fair trial, because any doubts about my destiny being to live in the North were cast aside forever.

Hughie and Phil, my two brothers, had previously gone to white fox country and had told me about this animal. They stopped by while I was still on the Slave River with stories about a remarkable territory waiting for trappers. Hughie said if I ever once got there and saw the white fox, wolf, and caribou trails that I'd forget about everything else and take up the chase.

In the spring of 1930, Hughie met me in Vancouver, knowing full well that I was looking for a way back to the trap lines. We weren't long in making plans and finding a way to come up with a grub stake. I was returning to my country and way of life. The deep drive within to do so had become more important than anything else on earth.

There is one definite thing about this life that I really believe: never do anything you'll be sorry for. If you do nothing wrong, you have nothing to worry about. My marriage was wrong. Moving to the South to live was a mistake. And so I made a decision to return to the land of the North. That was the correct fork in the trail for me and I've never been sorry I took it.

Over the years, I've been asked many times whether, if I had it all to do over again, I would come to the North to go trapping. My answer has always been the same: I'd get the best guldanged outfit I could get my hands on, and strike out for green country. Of course, I'd come back to the Barrens for trapping. It has always been in my blood and I couldn't get it out. Whatever a man's work is, trapping or what have you, he has to be happy while doing it. And nothing has ever come close to making me happier than running a trap line.

CHAPTER III

So it was in 1930. I'd run around in circles a few times and gotten lost a few times more. But this year marked the big turning point in my life that would enable me to enjoy the happiest years I've ever known. With the elimination of all possibility of living with my family in the South, my conscience was clear as I struck out for the Barrens seeking furs and whatever else it had to offer. And the animal about which I anticipated learning the most was the little white fox living out on the tundra.

Stories had filtered back to me on the Slave River all the time I was there concerning this mysterious little creature who spent his days out in the open feeding primarily on ptarmigan and lemmings. Someone told a story about a family of Eskimos who'd struck it rich with the white fox. Sometime around 1920, while traveling along the coast of the Arctic Ocean, they'd come upon a large, dead whale that had washed ashore during a storm and was well above the tide. The Eskimos were nomadic at that time and since the fur was prime and plentiful, they set up camp on the spot. They stayed right near the whale carcass and strung their traps around it. They remained there all winter long, and the fur kept coming in.

As the story goes, they ran their traps three or four times a day, never worrying about scaring the foxes away since the little critter experienced little or no fear of man. By spring, the Eskimos had accumulated a huge pile of pelts, the value of which approached thirty-thousand dollars.

When other trappers and I heard this story, it made our ambitions run high, to say the least. With my interest kindled and my enthusiasm running wild, it wasn't difficult to find enough courage to get beyond the tree line. In fact, that first fall on the Barrens, I could hardly wait until winter set in.

Perhaps the first man to tell me about white foxes had been Jack Hornby. I'd heard about Hornby when I first came into the country in 1920, and had met him in the early days down on the Slave River when he was on a journey south.

The legend of this man is almost unbelievable, but all the stories about what he did and how he did it were true. He wasn't a trapper or explorer; simply a Barrens wanderer. When he got the itch to travel, he'd just strike out, taking off for new country and wintering wherever he happened to be at the time that the season struck. He never took an outfit of any kind. He depended exclusively on the country for survival. Hornby was always in a state of depression and, much of the time, was about half dead.

I doubt that he ever owned a trap. At least I never heard from anybody that he ever got a bunch of foxes. Since he never took supplies in with him, he was in trouble whenever the caribou were missed. He told me he used to eat almost anything just to stay alive. After finding out more about his way of life, I had no reason to disbelieve his words.

Finally, I think in '26 or '27, old Mother Nature caught up with him and dealt a blow that cost Hornby his life. Along with two other fellows, he was found dead in a cabin on the Thelon River. Either the caribou missed them and they'd starved to death, or they'd died of scurvy.

Joe Nelson, another old-timer who arrived in during the twenties, came down to the south telling stories about the foxes. Nelson did some trapping and would strike out alone, headed for Artillery Lake on the edge of the Barrens and to the north of Great Slave Lake. In those days, that was wonderful fox country and Joe confirmed it with the fine reports he brought out of there.

A fascinating thing about Nelson was that the natives depended on him for news about the country he trapped. If he came back with reports that the caribou were out, or that the fox run was on, the natives would strike out. In those days, the natives were really good at pursuing their natural forms of making a living. In fact, many of them lived out there permanently, only coming out to sell or trade their furs. They don't live this way anymore.

Other early entrants to the country were the Peterson brothers, also arriving in the early twenties. They were Scandinavians and I don't believe Peterson was their real name—they adopted it because their own name was too difficult to say or write. They began trapping down on the Slave River above Fort Fitzgerald. Strong willed and determined, they pushed a little farther every year. The country just didn't seem big enough for them, so they kept going on and on.

Finally—it probably took them four or five years—they reached the end of the trees and the beginning of the Barrens: the Granite Falls area of the Thelon River. They made it one year ahead of my older brother, Hughie. On the way, one of them shot a caribou and then set traps around the remains. Every morning for six consecutive days they took a white fox from one of those traps. That told them all they needed to know. Those boys were trappers, and went after foxes in a big way. When they eventually came out in the spring, they had over 200 white fox pelts.

All of this news I was hearing about the white fox was firing my enthusiasm as I came north with Hughie headed for the Barrens for the first time. We had our stake of food and a good outfit. The desire to catch a fox was sending me toward my future home.

We came down the Peace and Slave Rivers to Great Slave Lake, and then across it. Yellowknife, now the Territorial capitol and a bustling, growing city containing nearly one-seventh of the Territories' residents, was only a small Indian village in 1930. Forty years would do things to that place.

It was July when we passed by there in a canoe on our way to the edge of the Barrens. We went up Great Slave Lake to McLeod Bay then northward to Aylmer and MacKay Lakes. Traveling the long water route and going into the Barrens over the portages was the hard way but, in the early days, the only way to get there. And I enjoyed it.

Following the trail Hughie and I took to get to the Barrens would probably inspire feelings of futility in the hearts of most men. Just the trip to the main camp in the Barrens would probably make many men who considered themselves tough and hard-working eager to return to a less strenuous way of life. The trip took well over a month and was extremely difficult and filled with hardships.. After all, we spent eight months of every year not only working at our job, but living with it, as well. But we weren't complainers, by nature. We seldom griped about the inconveniences we endured, instead passing them off and returning to them as topics of idle conversation. It was necessary that we maintain a sense of composure to overcome our troubles in times of discomfort. Indeed, I've always considered myself a proud man, and believed that proud men would sooner speak about happy times than troubled ones. This approach tends to prove a man's worth overall.

Probably the worst evils of spring and summer were the insects, which sometimes came upon us in swarms. Their tormenting ways had driven many less determined outdoorsmen scurrying for the safety of civilization, never to return to the wilderness. They made me wonder at times, but could never get me down. I just kept telling myself to accept the bugs because they were a part of the natural North. Anyway, I couldn't do much about them, and so fought them off as best I could.

Exposure to the various elements was another obvious source of misery, from time to time. This was inevitable, but most of us who spend our lives outside tolerated it. Besides, what was misery for some men was joy for others. Wind, waves, storms, blizzards—all of the things that nature threw out—were better than life in the city or on a farm. Much of the joy I felt was so only because of getting to know and understand the wilderness and nature. At an early age I learned the difference between bravery and foolishness. Taking no chances when the features of the day spelled trouble was a way of life I tried to follow closely. Patience on the part of the wilderness traveler is pretty critical.

Portaging eight months of supplies and equipment over the trails leading around river obstacles and between lakes was hard work. There were times when my back ached and my lungs felt like bursting—but the thought of a fox or wolf pelt quickly sent these annoyances away. The complaints were swallowed when happy thoughts took over. Most times, I met such arduous tasks with enthusiasm because they marked another step closer to the objective at hand.

Still, I wasn't spared when nature decided to deal from the bottom of the deck. A pretty common affliction on the frozen Barrens struck a blow to me that would affect me the rest of my days. Snow blindness is normally defined as a temporary blindness caused by the glare of the sun upon the surface of snow. But when the condition is applied to me, the word "temporary" should be omitted. This little dilemma cost me the sight of one eye my first year in the Barrens.

The type of snow blindness that took my eye was not the usual sun-glare variety. It first hit in mid-winter when the days were short. The sun was up only five or six hours a day and was a long way to the south. The snow blindness that afflicted me was caused by snow and sand, driven by strong winds.

During blizzards and ground drifts in the Barrens, the snow is like icicles or sleet. More accurately, it's like little balls of icy buckshot and, if a man doesn't protect his eyes, it'll play havoc with them. Carried by a strong wind, the snow stings when it hits your face. It affects dogs, too. From time to time, some of mine have gotten their eyes matted shut. Their eyes roll up in the sockets and turn red.

Anyway, the same thing happened to my eye. It became infected and hurt like hell. All winter long, I had headaches that nearly drove me out of my mind. Even though the pain was bad, I stuck the winter of 1930 out and never stopped trapping.

We left the Barrens at the end of the trapping season and headed south with our catch. I went to Edmonton in the spring and found a doctor. He gave my eye an examination, then carefully drained it. After cleaning it well, he indicated that it would probably be alright. But the next morning after I woke up, the darned thing looked as bad as ever. I returned to the doctor and, this time, received the bad news that it had to be removed without delay because the infection could spread and endanger the good eye.

The doctor put some kind of dope on it which sort of froze the socket and, as I sat right there in the chair talking to him, he took my eye out. Surprisingly, the operation hurt very little. The bloody thing had caused me so much misery that I was glad to get rid of it. After losing the eye, I always had a tendency to track to the right when breaking trail. Perhaps I was trying to compensate for my blind side.

I never suffered any bitterness or remorse toward the land and its character because of my misfortune. I felt too strongly about the North and the way of life it offered to slander it. I accepted the accident as just one of those things that might happen during the course of events that go into the making of a lifetime. Hell, I didn't give up. Not one inch.

Losing that eye never changed my way of living or caused any delay in my returning to trap the Barrens animals. There was a job to be done and all men have to work at what they like and do best. My coffers weren't lined with gold, so I had to earn a living. I bought a glass eye and headed North in the late summer. Glass eyes were the only kind of replacements available then. I had to get a new one from time to time because I'd get careless and drop it. Usually it rolled up against something hard and broke. When they came out with a plastic eye, I bought one of those. But after a plastic eye gets some age on it, rough spots develop and the eye socket gets scratched.

Once, I lost a prosthetic eye in a snowdrift out in the Barrens during early winter. Funny how it happened, too. After washing the dishes on this particular evening, I opened the tent flap and tossed the water out in the snow. At the time, I noticed nothing out of the ordinary at all. Besides, it had been quite a tough day and I was in a hurry to get to bed.

The next morning after getting up, I still wasn't aware that my eye was missing. You don't really notice its absence unless you check for it. But when I went out into the air, a cold draft set me straight. I reached for my face and immediately discovered the missing part. I knew bloody well that the thing had been in the socket when I'd gone inside the night before; it had to be lost somewhere inside the tent, which was only a seven by nine.

I went inside to look for it. After thinking about it, I decided it had to have fallen out when I'd tossed the dish water out the door the night before. I went outside and looked all over for it in the snow, but couldn't find a thing. I did a final sweep of the tent, but came up with the same empty-handed result. By then, it was getting late and I had to travel. Figuring the eye couldn't go very far and that I'd find it later, I hitched the dogs up and struck out for my line.

The next spring, I happened to be back at this out-camp just after the top crust had thawed slightly and chanced to glance down at the snow by the door of the tent. There the bloody eye was, looking right up at me from the snowdrift. By the Jesus, it sure looked odd and was funny, too. Anyway, I got my eye back after a cold stare.

During the next couple of years, Hughie and I trapped in this same general area west of Aylmer Lake in the region of Backs Lake. The loss of my eye had quickly become an insignificant thing of the past. Fur was plentiful, trapping was successful, and our style of life was absolutely ideal. We were alone, enjoying our freedom, and worrying about few things beyond surviving and catching fur. Caribou had been plentiful, too, so wolves were also abundant.

In the summer, we'd get our outfits together along with a couple of good dog teams, then strike out for Fort Reliance at the north end of Great Slave Lake. From there, we'd leave early, intending to arrive at our main camp in the Barrens by September. We depended on the caribou and killed enough for a winter's supply of meat for ourselves and our dogs. These were cached for safe keeping and dug out as they were needed. Next, we'd work on our wood supply, stockpiling enough at our main camp to last all winter. This wasn't hard work, but an enjoyable task that passed the time while we waited for the trapping season to commence.

Once the fur became prime in December, we put our traps to work until spring approached. We usually went south in April to sell our catch and relax for two or three months. Then, once this stretch of leisure came to an end, it was back to the Barrens again.

You know, I used to get scared every time I went outside the North in the spring to sell my furs. Somehow, I was afraid something drastic would occur and that I would be unable to return to the land I loved. The South wasn't all bad, but I was never comfortable there after tasting the vast atmosphere of the far North. I was drunk with its beauty and mysticism.

Mind you, the North wasn't a bed of roses all the time. I experienced several bad times there, like the one when I fell through the ice when the temperature was thirty below zero. I should never have been out on the lake and knew it at the time, but I'd replaced caution with carelessness and became reckless. Instead of following the shore line—the sure way—I'd foolishly decided to cut across this narrow bay.

Well, I went through. When it happened, I came out faster than a jack-in-a-box. After getting to shore, my search for dry matches was mercifully brief, as the ones in my shirt pocket were still dry, so a fire was soon crackling in the cold air. My clothes eventually nearly dried out, and I ran all the way back to the cabin a much wiser man for my actions, swearing never to pull that trick again. With this, I was one up on perhaps the greatest danger in the North, and figured I was lucky to be alive.

Ice is treacherous. Up here, you just don't go out on it until after it cracks, no matter how cold it's been. After three or four days of hard freezing, it'll finally crack. Water comes up through the cracks, runs out on the ice and freezes. Only then is it safe to travel over. From that time on, I watched for this to happen and never got caught again.

I studied the habits and weaknesses of the white fox and my ability to catch this little ball of fur improved with each season I spent on the tundra. Through a lot of years in the bush, I had become well acquainted with nature's animals and their habits. More than that, my natural senses had become sharper and I could better read the outdoors. Observation is the key to learning and understanding and I had plenty of time for that. Trapping had become an art and a science, and success with it depended upon man's ability to match wits with the wily animals on their terms at their place of residence, the tundra.

I learned early that the white fox wasn't found everywhere. He flourished in cycles with the lemming and stayed close to where this little rodent lived. When lemmings were resident in an area, fox tracks appeared everywhere, crossing both ways over the lemmings who had tunneled below. This is where I set my traps.

To me, the lemming is a handsome little animal who, at about a foot in length, looks rather like a southern weasel. During the winter, they change from their summer brown to pure white, except for their backs which are somewhat gray. They're an interesting little animal that are found around the Barrens in three or four varieties.

Lemmings eat tons of vegetation, mostly roots and grasses. Much like moles, the little fellows tunnel under it down to the rocks and sand. Foxes listen for this digging, and pounce when they hear it. Once in a while, I would see a blotch of dark-looking stuff where a fox had dug one out and eaten it. Sometimes during a winter mild spell—still below zero, maybe ten or fifteen degrees—lemmings would come out on the snow. I've seen their tracks criss-crossing on a side hill. They don't jump like a mouse, but walk like a fox.

When the caribou were around, wolves were seldom far away, and I pursued these predators with a vengeance. The cunning of this animal has been described since the beginning of history and, in many cases, the words are true. Nevertheless, I got my share of them after once learning some of their weaknesses.

Occasionally the tough, fearless wolverine ventures out beyond the tree line into the Barrens, either on a feeding circle or in pursuit of the devilish deeds his nature demands. Whenever one showed up around me, he wasn't safe for very long. Since he was a regular twister around a trap line, a few foxes always were lost. The wise one would eventually have to be caught. I would go after him with everything in my power until I got a trap over his leg. No wolverine was ever safe around me because I wouldn't let him be.

It wasn't until well into the third winter on the Barrens that I felt totally sure of myself. By then, I'd pretty well mastered the essentials for making a life for myself in the North's brutal conditions. Each year was a new learning experience and I made progress in many areas. Even out there, I maintained the belief that a man should never go against his own nature, and I put this philosophy into strict personal practice. I felt a breadth of freedom I'd never known before.

I had surely found the place on earth for which I was intended. My dogs—together with the trails of caribou, wolves, and foxes—were the only companions I needed for surviving the long winter months. I found peace of mind inside my skin tent as the cold Arctic winds whistled in defiance. I was the intruding stranger and happily conformed to the ways of the land.

I suppose I survived because I was one of a rare breed of men called Barrens trappers. Those of us who braved that great white expanse had feelings for it that could never be understood by the man on the outside. I never dreaded the coming of September when I left the civilized world behind. In fact, I counted the days until the journey back to Fort Reliance would begin. Occasionally, I'd even get there a couple of weeks early, just to swap yarns with the other trappers.

Even though Fort Reliance was the big gathering place for several trappers in the fall, our spells spent there as a group would mark the last time we'd see each other until the following year. A close neighbor out on the trap lines in the Barrens was a fellow trapper who was maybe twenty-five or fifty miles away by dog team. Social visits out there were few and far between. When we pulled out for our trapping territories, there were never any greater individualists anywhere.

There was an understanding between us when it came to territorial boundaries. These agreements were like unwritten laws and were seldom violated. We knew it was a big country with room for us all. Thankfully, we were men of integrity with a common set of goals and cooperated to achieve them. We had enough skill in our professions to make it on our own and didn't need another man's fur. When the fur was plentiful in a man's territory, he made a good catch.

Neither did we envy a man who enjoyed success with his catches. He was congratulated. Similarly, if someone had a bad year or missed the run of fur, he wasn't put down. We were always exchanging ideas and learned from each other. There was a mutual affection and a bond among us filled with humor and jest. Seldom did our actions and deeds go beyond a point of no return. Out in the cold, treeless regions, you just never knew when you might need one of the few people around.

I went out of my way to cooperate with my fellow trappers and didn't speak badly of any man I recognized as an equal. It wasn't difficult to find good points about all of these men, and I loved to share tales of their great accomplishments. I suppose all of us contained a little bit of each trapper we'd been in contact with.

It seemed as though all the trappers were men with a great many friends both within the circle he walked and on the outside as well. Living alone for so many months caused a man to cherish, and hold deeply to all friendships. On the outside it was easy to make friends because we were considered adventurers and people eagerly awaited the stories coming from the cold North—a place forbidden to them by their choice.

Most of us had a character marked with a distinct stamp of self-discipline without which we could not have lived to return from our trap lines. Each of us had his own nature and habits, which made for a vast cross-section of individuality. The air of freedom that captivated the trappers was impossible to duplicate in other avenues of life. We were beholden to nothing beyond the basic laws of the Territories and the rules by which we had to live to exist.

I've long felt sorry for other men who lack the peace of mind we felt while doing our work. We had a passion for our work and understood the risks involved. We never worried when things took a bad turn, because there would always be next year. Anyway, even if we missed the fur, there was usually more left over than there would've been working for the other man on the outside. It surely didn't take much to live out there. Besides, what would a trapper do with all those things that caused most people so much trouble?

All my life, I've tried to be a good example of a true Barrens trapper. That image is important to me. I'd like to believe that my way of life has been meaningful, and that the trapping in which I've engaged has contributed to the overall importance of this fine country. I'd also like to think that one reason this is a fine country is the people who opened it up and later started a trend to populate it.

CHAPTER IV

There are times and happenings in the lives of all men that would be better forgotten and most certainly never duplicated. The trapping season of 1933-34 brought with it a series of circumstances I'll never want to face again.

The trip in from the South with our outfits was typically uneventful, and offered a welcome change from the civilization with which we'd been bumping elbows all summer. As usual, we pulled into "Trapperville" at Fort Reliance toward the end of August. Although this journey to the north end of Great Slave Lake was long in both time and distance, we seldom arrived worn out. Soon after issuing our greetings and exchanging a few tales with fellow trappers, we became impatient to get to our camps in the Barrens and get on with our business.

On this winter, my brother Phil intended to trap the Whitefish Lake country while Hughie and I were heading the other way to winter again at Backs Lake, to the east. This area had been rewarding for us in the past, so we felt no need to search for new country.

Hughie had some physical problems, and hadn't been feeling well over the last part of the summer, so he'd decided to fly in later to meet me with part of the equipment and most of our food. So it was that, in early September, I pulled out with the rest of the trappers expecting to hook up with Hughie a few weeks later at the main camp we shared. Hughie had wanted to take the easy way in and it made no difference to me, but I wondered at the time if he was really sick or just dreaded going in with us because we took the hard way. We went by canoe portaging and packing in, and it was danged tough work.

But the work really began when we came to Pikes Portage at the end of McLeod Bay. This portage showed the way three miles overland to a series of lakes that ultimately led to Artillery Lake. The area between Pikes Portage and Artillery Lake at that time was hard work.

Several trappers went out over the portages at the same time I did, but Artillery Lake was the jumping-off place. From here, we would branch out into several different areas. Some trapped south and east into the Lynx Lake and Beaverhill Lake areas; others ran their lines around the Hanbury River and on into the Thelon River country; some of us went north and west to Clinton-Colden Lake, Aylmer Lake, and Lac Degras; and some stragglers stayed around Artillery Lake, which was also good country.

After we had made it to Pikes Portage and gone through the rough stuff to get to Artillery Lake, the going should have been pretty routine. We had to cross Artillery Lake to get to the west shore. The place of crossing was only about four miles wide, but it could become pretty danged rough when the wind was coming directly down the lake from the northeast. On that day, the inshore wind was blowing hard from this direction and the water was quite wild. It rolled and tumbled toward the west shore where we had to go, and the conditions were undoubtedly risky.

One look at the scene, and I knew crossing then wasn't for me. I was against it all the way. The correct move at that time would've been to pitch the tent on the spot and wait her out. After all, when the wind is blowing hard, a man shouldn't be out on a lake. It's totally unreasonable to take foolish chances when your life and supplies might be risked. Out in the wilderness and without immediate access to assistance, a man learns to watch his step. But in the process of learning many of us do some damned fool things. I was foolish this time.

George Magrum, a fellow trapper and a good one, was along on this trip. Since he was trapping the same area Hughie and I trapped, we decided to travel out to our camps together. His main camp was in a location twenty-five miles from ours, making us neighbors (such as neighbors go in the Barrens).

Magrum was impatient to strike out and believed the water wasn't rough enough to hold us back. He told me the crossing would be simple enough for him, but his outboard motor was sputtering and wouldn't run.

He continued to say encouraging things about the weather and conditions, and eventually talked me into taking off. I left a line back to his canoe to tow him behind.

I bloody well knew better than to try this, but a man sometimes allows himself to get talked into things against his better judgment. Too many times, in these cases, the results turn out to be about equal to the original fears—and this was to be no exception.

By my standards, Magrum's canoe was unsafe because it had been altered. The canoe had originally been a couple of feet longer, but he had cut the pointed back end off and boarded it up flat so a motor could be bracketed on. After that, it never seemed to balance a heavy load real well.

It was loaded down heavily with gear on this trip, since the plan was for a one-time crossing only. My canoe wasn't as heavily loaded, mostly carrying hardware and my dog team. With not too much trouble, everything was put in order and we embarked for the west shore with Magrum and his son perched way up high on top of their outfit. They were in a precarious position, especially since neither one of them could swim a stroke.

The farther we went, the higher the waves became. The swells got bigger and closer together and, by the time, we were approaching the other shore, she was blowing a regular gale and the water was much rougher than it had been at our embarking point. It became so bad that I decided something drastic had to be done and done doggone quick. If not, we'd surely swamp well out in deep water. I didn't know what else to do, so I cut Magrum's canoe loose just off the shores of Crystal Island. I still firmly believe if I hadn't cut them loose when I did, all of us would have drowned right then and there.

Anyway, as soon as the tow rope was cut, their canoe turned sideways and swamped immediately. Since their load was too much for the canoe and their high perches atop the load caused it to be top heavy, over they went. When I looked back at them, the only thing I saw were two heads. This really scared me. Even though the bloody canoe was loaded with food, heavy stuff like flour and dry fruit, it was of little concern compared with the lives of the two Magrums. The canoe was already on the bottom. The only thing holding them up was the air captured inside their tent and tarps.

With as much haste as the circumstance would permit, I turned my canoe around to head for the rescue. All this time I was in trouble, too, and getting to them was difficult. When I did, they wanted to climb over the side and into my canoe. I told them to forget about doing that because we'd all go under. If they'd tried to crawl over the top in that kind of water, they would've sunk me, too.

Instead, I directed them to grab onto my canoe and hang on until we could reach the safety of shore. Neither one could swim, so it can be stated emphatically that they hung on tightly. I headed to shore knowing I couldn't be too careful. Sure enough, just as we got near shore, my bloody canoe took on one wave splash too many and also swamped. This was fully expected, and so I wasn't completely unprepared. The shock of cold water would've been too much, but luckily the depth was only up to our chests. We stood there with everything in a state of confusion.

I'd prepared in advance for just this calamity, and a quick concern for my dogs was all that saved them from drowning. They had been chained apart inside the canoe to keep them from running around and getting tangled up in a big fight. Many trappers learned the hard way that a canoe is no place for a dog fight. My feet no sooner struck bottom than I had my knife out and was moving from dog to dog, cutting collars. Everyone was saved, and swam quickly to safety. I got to the last one just in time. You've never seen a dog team so happy to escape a place.

After freeing the dogs, I looked around Artillery Lake and discovered a weird scene. The surface of the tumbling water body looked like the aftermath of a shipwreck, with items from my outfit floating around everywhere. There was no humor in the situation, as the water and wind were chilling us to the bone. And the seven or eight inches of new snow on the ground at the time compounded our problems as we trudged to shore. Absolutely nothing was dry. Everything and everyone were completely soaked. When we hit the lake shore, we kept moving, all running around to try to get warm and gain a little comfort.

Somehow, Magrum managed to come up with some kind of a lighter and, before long, had it working. That was fortunate, because every match among us was drenched beyond value. We proceeded to build a huge fire out of ground spruce and anything else that would burn. Throughout the drying period, we maintained a silence, each man pondering his own predicament and settling on the next step to improve the situation. Each knew the problem was a serious one, but also realized it wasn't hopeless.

After thawing out and drying our clothes, we pitched a tent we'd managed to salvage. We tried to sleep, even though there were no beds. There was also no cooking utensils and nothing to cook. What a time! Somehow, we made it through the night without freezing, although there was very little sleeping.

By morning, the gale had died down, enabling us to make an attempt at retrieving some of our sunken supplies. My canoe was easy to reach and, with little trouble, was surfaced with everything pretty well intact. Magrum was pretty danged lucky, too. After some maneuvering, we managed to get a rope down to his sunken canoe and somehow hooked the flywheel of his outboard motor. At a depth of thirty feet, it was more blind luck than skill that made this possible. Anyway, we pulled the canoe to the surface and were able to salvage most of his equipment. We managed to save most of his food, too, even part of the flour and dried fruit. Our spirits improved remarkably after that, and our thoughts became more positive.

I'm certain George Magrum never forgot that incident. I'm sure he always blamed me for swamping them and sinking their gear, but from where I stood, there had been no choice. I felt then, and still do, that if I hadn't cut that rope when I did, we'd all have drowned before making it to the shore. Still, every time I've ever thought of that incident, I've felt remorse and sadness. Situations dictate to man. Everyone does what he thinks is best at the time, and when you get into a spot where something has to be done, we all make our picks. I know very well that if the same situation were to arise again, I'd react the same way.

Things were finally pretty well straightened out, so we continued on our journey toward our trapping grounds. Our troubles over, the old enthusiasm once again made itself felt as we pushed onward. I told Magrum that we'd better stop somewhere around Clinton-Colden Lake to either net some fish or kill a few caribou. He disagreed, saying meat wouldn't be a problem if we pushed on. This difference of opinion, however slight it seemed at the time, again demonstrates why I always liked to be alone: I get to work by myself. A man can get into too much danged trouble when he starts listening excessively to other men.

The canoe had long since been cached and we traveled by dog team. The snow was plentiful and just right for making good time. At Clinton-Colden Lake, we didn't slow down as I had wanted to, and for two days beyond the lake, the dogs kept steadily on the move. And then all at once, we crossed another set of toboggan tracks, which surprised us to no end. Upon stopping, Magrum wondered aloud at who the hell else was out here.

Looking around carefully and studying things closely, I came to my senses. By the Jesus, I realized those darned tracks belonged to us. We'd been traveling in circles and had finally crossed our own trail. When a man gets in the kind of hurry we were in, he without doubt starts making careless mistakes. We weren't greenhorns, and this kind of behavior woke us up and kept us on our toes.

Having come to our senses, the compass again became our guide and squared us away. We set out on our corrected course with the dog teams suffering from hunger. Our dog feed had been exhausted. Our circular jaunt combined with our failure to stop at Clinton-Colden Lake to hunt and fish had been enough to spell doom to our reserve dog food. We had to find something for the dogs to eat before very long.

Continuing on our route, we came to a small lake and I raised my head to glance across it. Over near the far shore stood a beautiful sight to a hungry trapper: a big bull caribou. Right away, I talked to my dogs and we set off after him in a whisker. When I got close enough to shoot him, I pulled on the main line, gave it a couple of turns around the upright, and dropped to one knee. I had taken the rifle out when I'd first spotted the caribou. Now I lined it up and let a shot go. The bull dropped in his tracks, and I was immediately a very happy hunter.

Before skinning the animal, I looked him over carefully and discovered he was a cripple. He had something wrong with the joint in one of his forelegs, but was still in very good shape. He had, like all well-fed caribou in the early fall, a row of fat down his back. This helps to make caribou meat the best meat on earth. A man doesn't even have to chew it; it just melts in the mouth. I was lucky to get this big fellow. If the wolves hadn't gone on out after the main herd, they would've had him.

The caribou provided a little food for us and a break for the dogs, but it didn't go very far after spreading it around. Each of us took a share and then split up, figuring we'd have a better chance at running into some caribou on our own.

I hunted hard all the way to Magrum's main camp, but the caribou had already passed to the south. By then, my dog feed had run out again. Magrum hadn't gotten there yet; somehow I'd beaten him in. There wasn't any meat around his place and the dogs were in a seriously hungry state. Since there was no other choice, a dog had to be killed to feed the others. I sacrificed a female and cooked it enabling the others to survive. No man liked to do this, but there were no other alternatives.

A short while later, Magrum finally showed up. He had really extended himself and hunted hard, but had suffered the same misfortune I had. He'd found no caribou, either. We agreed that we'd clearly missed the caribou migration. They had either gone south earlier in the fall, or had bypassed the area completely. The caribou are unpredictable and, one year, will mysteriously be absent from a route they've used for years. You can never be absolutely sure where they'll pass because, when they strike out from the Barrens toward the wintering grounds in the trees, they take a straight bloody line and stay on it. I never found out what happened this time, but knew things were getting pretty sticky for us. We were in desperate need of meat.

The day after Magrum arrived, we made a hunting trip, but returned to the camp empty-handed. The following morning, Magrum looked out the door and saw a herd of caribou, maybe a hundred, coming directly toward his camp. Right at it, mind you. He and his son dressed quickly, grabbed their rifles, and struck out for them.

Another trapper had spotted them well to the north and was trailing them from the other side. Somehow, there was a big mix-up and the caribou spooked and missing everyone. The darned things shied away from Magrum and not a caribou was shot.

Magrum and his son came back to the cabin and we fixed a breakfast and ate it silently. Everyone was pretty danged blue. There had been meat right before our eyes and it escaped without us getting any of it. What had been a difficult situation was now compounded by a feeling of despair.

While eating, I paid little attention to the others and had very little to say. My head was occupied with ways out of our trouble. When we finished our meal, I told the others I was going out by myself. Every herd of caribou has some stragglers, and I was hoping this one would be no different. By going alone, I could let my own judgment rule and possibly eliminate the kinds of problems we'd run up against earlier. I picked up my rifle—in those days I carried a half-magazine 30-30 Winchester, and a danged good one, too—and with my snowshoes in place, struck out. The others decided to go out, too, but in a different direction.

I reached a hill a short distance away which I crossed to a ridge heading down the other side. When I got there, to my surprise, I caught a glimpse of some movement which turned out to be caribou. Five big bulls were coming directly toward me. I quickly dropped to my belly and waited, scarcely believing my eyes and wondering if my luck would hold out. I dug up some dry moss and put it beside my head, and then took some extra shells out of my pocket and laid them on the pallet of moss. The caribou stayed on course, coming right for me.

All at once, a shot sounding no more than a quarter mile away rang out, and it caused my heart to drop into my mukluks. I felt sure that the hunt was over and that failure was on the return. And then, about the time I was feeling confident that my chances were gone, a big bull came charging right at me like the mill-tails of hell. He was so danged close, I didn't even aim my rifle, but picked it up and plugged him, point blank. He was dead before hitting the ground. Knowing I had him in the bag, I turned my attention to the others. Somebody up above was with me because they hadn't faltered and were still coming my way. Not wavering an inch, they came nearer and nearer. Finally, they were just right and I went after them. I killed every bloody one of them without missing a shot. Anybody who has ever hunted caribou knows this is not an unusual happening. But to have it occur at such a needy time made me feel very good.

Completely elated with the situation and devilish about my success, I went to work skinning and gutting the caribou. I wiped and cleaned all the blood and other evidence from my hands and clothing, then walked back to camp. There was no response from the others when I walked into the cabin to indicate that they were aware of my shooting. At this point, we were all accustomed to failing hunts, and so not a word about success or failure came up.

Breaking the ice after a ten-minute silence, I asked about their hunt. They sadly disclosed the near miss of a lone bull at long range, and made no mention of having heard my shooting. Unable to withhold the good news any longer, I informed them that we now had enough meat for a few days. Their shoulders squared and a new twinkle of optimism appeared in their eyes. A spark of hope now dented the gloom that had surrounded us.

Even though the mood had reversed from one of discouragement to one of anticipation, not a great deal was said among us. Perhaps that was our way of taking things pretty well as they happened. Calmness and lack of excitement were simply our nature, notwithstanding whether feelings of happiness or sadness were prominent at the time. Of course, the attitude of the moment had changed. The immediate problems were over, but six caribou weren't going to provide enough meat for all of us and our dogs for very long.

In short order, I prepared to leave Magrum's camp and head toward my own camp in a day or two. I carried only enough meat to last until arriving at the camp where Hughie was to be waiting with supplies. On the seventeenth of September, I pulled into camp expecting to be greeted by my brother. Was I in for a surprise! Hughie wasn't there and, what's more, hadn't been there all fall. I was worried because that just wasn't like Hughie. I didn't know what might have happened to him.

So there I was in one helluva fix with no dog food and nothing to eat myself. Since the meat from Magrum's had been consumed, all the sleigh contained was hardware. The food was to have been brought in with Hughie on the plane.

I took the .22 from the sleigh and started to poke around for some kind of small game. I located a couple of Arctic hares and was luckily able to shoot them both. One of my dogs, White Brandy, grabbed one of the rabbits and absolutely refused to release it. That big white bugger wasn't like that, either. He was typically mild mannered like the pet he'd become over the years. But along with the rest of the team, he'd been working hard and his hunger was approaching starvation.

The impossibility of the situation dictated the next move, which was to return to Magrum's camp immediately. There was little chance of surviving at my camp without caribou meat or Hughie's supplies.

Back at Magrum's, the six caribou I'd shot were nearly gone. Soon, a meat shortage would be upon us again. Still, a couple of mornings later, things looked somewhat improved when we spotted four caribou across the Lockhart River on Aylmer Lake. Although the lakes were frozen and safe for traveling, the river ice didn't look right yet. But Magrum didn't want to wait around and, my efforts to talk him out of it given that the ice on the river was too thin to cross with confidence notwithstanding, hurried after them right away. He was desperate for meat and knew that another opportunity might not present itself. So he took the chance.

Having laid two long poles apart on the ice, he lay down on his belly and slid across the river, using the poles for support. The ice sagged and cracked, but those two poles spread his weight over enough area to keep him from going through. Not only was he able to cross the river safely, but to kill all four caribou and drag the meat back using the same crossing technique as before. Magrum demonstrated a lot of nerve during that risky piece of business—more, probably than I could have. I doubt that all the tea in China would have pushed me across.

After all the meat had been hauled to Magrum's camp, I heaped a load on my sleigh and headed back to my camp, hoping that Hughie would have arrived by this time. No luck! When I got there, everything was as it had been before: vacant. The load of meat from Magrum's kill wouldn't last very long, so I went hunting nearly every day.

Once, I spotted a lone bull well north of camp and was able to get close enough to plug him. While I was skinning him, another trapper arrived to say he'd heard the shooting. Also out of dog food and in bad shape, he asked for some of the meat. Even though I was short, I couldn't refuse him. It was hard to part with that meat, but I did.

Somehow I managed to stay around until November, expecting Hughie to show up every day. But there was no sign of him and I'd waited as long as I possibly could. The time had come to leave, so the other trapper hooked up with me and away we went toward the south. When no more than thirty miles were behind us, we found caribou everywhere.

The country was by no means strange because this was on the route we'd taken out the last spring. A big range of hills lay to the west and I knew exactly where we were. A short distance away I saw a mound, or rather three of them that looked out of place. I paid them little attention as I was in a hurry to make camp and start hunting a supply of meat.

As soon as camp was in order, we lit into the herd with a flurry. After the shooting was over, two shots rang out in the distance. This was strange and unexpected, but also familiar. That was the signal Hughie and I used in case there was ever any trouble or need to contact the other.

Since we were poaching on the Yellowknife Reserve, the other trapper was dubious about answering the shots. He worried that it could be a stranger and was afraid we might get into a jam. Telling him it was too late since our other shots had been heard anyway, I let a couple of answering shots go, knowing it was likely Hughie and not wanting to chance missing him. Killing the caribou where we did was undoubtedly wrong, but this was a food emergency. This area is no longer a reserve and it no longer matters.

Soon after firing the two shots, we began cleaning the caribou. All at once I saw from the distance a fellow approaching. Before long I could make him out. Boy, was I tickled, because it was Hughie. I asked him where the hell he'd been all fall. He seemed surprised that I hadn't opened the mounds and said he'd been here since September.

Then I remembered those extra mounds to which I hadn't given more than a passing thought because of the presence of caribou. If I had looked in them, the whole story would have been exposed. They were caribou caches in the top of which he'd placed tin cans containing notes describing everything that had happened.

As it turned out, Hughie had just as bad a time getting out there as I had. His flight had gone out as scheduled and brought him to the area before Magrum and I had arrived. On the way, however, they had a helluva time with a bad snow storm and got lost in the air.

Punch Dickens, the pilot, took a chance, guessing at the intended location and put the plane down on a safe landing. They had flown around for quite some time at this point, and were low on fuel. Dickens knew he had to do something. After landing, Hughie wasn't sure where he was, but wasn't overly worried because he felt confident he was fairly close to where he wanted to be. Knowing he'd be all right because there was plenty of grub and a good team of dogs, he unloaded everything right there.

But when the pilot took off, Hughie really got screwed up in a hurry. The doggone plane took a course straight north, with Hughie listening in disbelief until the drone of the engine was beyond hearing. Since the route back for the plane should've been straight south, Hughie was left in doubt about his orientation. And nothing went right for him after that. Things were really twisted around because of that doggone nagging doubt.

Funny thing about that incident, too. The pilot actually was lost and got himself into trouble. After coming to his senses, he had gone so far north that he burned too much gas. He'd turned around and started heading south, but never made it back to McLeod Bay. He'd run out of gas and had to land the plane inland, somewhere north of the bay. Although unhurt, he was in a fix and wasn't found for eleven days. He was danged uncomfortable by the time Matt Berry discovered him and brought him out.

All this time, Hughie and I had been in bad shape. He had all the grub while I had all the hardware and traps. Neither of us could function without the other. First thing Hughie wanted to know was where in the world we were.

When I told him we were near Backs Lake thirty miles south of our main camp, all he could do was shake his head in disbelief. Although absolutely no stranger to these parts, he had been lost. He hadn't sat around idle, but had gone out several times trying to find something out.

When first landing, he had known things weren't exactly right, but didn't worry. He had cached everything carefully before striking out to do some scouting. He pack-sacked out, but wasn't able to recognize anything, and failed to pick up even a single landmark. Hughie knew this country, too. But somehow things looked different to him. No matter how hard he tried, he simply could not get his bearings, and so finally called a temporary halt to the searching. A big herd of caribou was staying close around him, so he started shooting a supply of meat. He downed and cached ninety in a short time. While he was at it, he observed an abundance of fox and wolf.

Somehow he had nearly lost his life while trying to get untracked. He'd gone by sled and dog team out on a lake and it hadn't been frozen hard enough. He fell through the ice with his dogs, some of whom drowned. Luckily, he managed to pull himself out and get to safety. He had enough dogs left for two teams, but some were green pups and not yet ready for the harness.

Now having enough meat for the three of us, we went back to our camps and started thinking about our business. Hughie and I made up one good team from the older dogs and started trapping. We each had our own lines, and alternated the team to run them.

Everything quickly fell into place after Hughie and I met. The season was upon us and there was little time to idle away. I was unaware at the time, but Magrum was a little put out with me. When this other trapper and I had headed south to come out, I had left a note in one of Magrum's out-camps to explain this plan and to say that, if I didn't find any caribou, I would return to Fort Reliance. Since I did find a caribou herd, he figured I should have returned to his camp to tell him about it. But I thought he'd either find meat where he was, or head for Fort Reliance on my trail and run into me. He just barely got enough meat to get by, by sticking it out all the way. He not only didn't leave, but made a good catch, as well.

The scant population of the winter Barren Land consisted of only the few trappers with determination enough to challenge the harsh weather. With our camps and trap lines so far apart and travel limited to dog teams, precise communication was unlikely and some misunderstandings were possible. Thankfully, our survival and success made us look ever to the future. The better things ahead preoccupied us, and we didn't fret about misgivings of the past.

A man unable to cope with small annoyances didn't belong in the Barrens. Even though a few instances arose which could have thrown a couple of us at each other's throats, judgment ruled, and a face-off was avoided. To this day, there is no one I respect more than George Magrum. He is a fine man and one of the best trappers to ever hit the Barrens.

CHAPTER V

Throughout the years, I've seen several men come into the North Country with stars in their eyes expecting a world of enchantment and gentleness. They were met with a rude awakening, and lasted but a short time before returning south with their wild dreams punched full of holes. A life spent following a dog team over the trap lines is marked with a wide variety of extremes from both a mental and a physical standpoint.

Perhaps because I've been out here so long, I've gotten the idea of what it takes to survive and enjoy this way of life. I believe the human male, before he becomes enough of a man to be called a man, has to be in control of a wide variety of mental and physical attributes that distinguish him from the unpredictability of youth. Probably one of the most important characteristics is his ability to endure hardship and suffering without giving way under the strain.

Determination—or, in my case, plain bull-headedness—is necessary for standing up to opposition when a man's objectives and very life are threatened. Sometimes, it takes all the gameness a man can muster for fighting back and not quitting when the odds are seemingly impossible.

Up to this point, the winter of 1933 had been testing the man in me. I still had my spirit, even though my outfit had sunk; I'd missed the caribou, causing the dogs and I a couple months of hunger; and my brother had been missing for three months with the supplies. After Hughie showed up, bull-headedness still wouldn't let me quit even with the meat thirty miles from camp, only one dog team for the two of us, and the trapping season well along before we got into it.

Ultimately, our strength of purpose prevailed and things reached a satisfactory plane of acceptance for us. Fur was coming in and, by the time a new year rolled around, it looked like a fine catch was on the way. By then, the confusion of the fall was cast into the background and the shadow of misfortune was now aglow with success. January and February were fine months, and March started the same way.

But this was all to end abruptly for me, and the trying events of the fall would take on a degree of very small significance by comparison to what lay ahead. The extent of my bull-headedness was to be put to the test in a situation of brutal proportions. I was about to find myself hopelessly lost, with my life directly dependent upon my ability to stay awake. To fall asleep without a shelter in the frigid Barren Land was a one-time/last-time occurrence after fatigue and hunger assumed control of a man's being.

It was well into March when George Magrum came all the way over to pay us a visit from his camp twenty-five miles away on Aylmer Lake. Visits in those parts were few and far between, so we stuck around the camp visiting for two or three days. We had several experiences to exchange, and that made time pass rapidly.

Toward the end of March in the Barrens, the weather gets pretty rough. Maybe some people don't agree with this piece of trapper weather forecasting, but we soon learned that sometime during the change in the moon, there'd be a storm. Since this early-spring storm was usually a bad one, we always tried to avoid getting caught away from our main camp at this time of month.

Because of our dog shortage, Hughie and I alternated our one veteran team on our lines, and it was my turn to go out. His line extended a different direction than mine which ran two-and-a-half days north to what we called Icy River beyond Outram Lakes, between Aylmer Lake and MacKay Lake. To guarantee that I got back before the end of the month, I told Hughie it was time for me to strike out over my line.

I reached my first out-camp on schedule the next evening, and the trip was progressing routinely and uneventfully, with only the usual ground drift marring it. The next evening while approaching my second out-camp, I spotted some caribou pawing around close to it and managed to kill four or five. Since the season was drawing to a close, this was all the meat I'd need at this out-camp.

The next morning before hitching up the dogs to run the half-day line, I decided to leave one dog tied up at the camp. Since I was to return and sleep at the camp that night, there was no reason to carry a big load and the dog wasn't needed. Somehow there wasn't a chain to be found, so I tied him with a small rope. If he got loose, the pile of caribou meat would keep him from running away. That bugger was part hound, and wouldn't leave that meat as long as a sinew was left.

There was no reason to take my eiderdown or a tea pail on such a short trip. About the only extra item I took along was an ax—a meat ax, of all things, which wouldn't cut anything. The usual few traps, a snow shovel, and my rifle were everything on the toboggan. A few cartridges and some matches were in my pockets. Since I was well-dressed in caribou skin garments with woolen unders, the cold wasn't a bother. With everything set for a short half-day run, I struck out.

Most of the time it's impossible to see the old trail ahead because nearly every day there's a ground drift that covers it. It's up to the lead dog to keep you on course. At that time, we were driving a good lead dog who was in excellent trail shape. If the bugger had a weakness, it was his craze for caribou. It was nothing for him to see caribou in the distance and leave the trail. He was a stubborn cuss and when he took off, that was it. Even though the caribou had been scarce early in the fall, a few had shown up on and off all winter.

On this day, with the end of the half-day line nearly in sight—and with no more than a mile to go—a cow and calf ran across the trail directly in front of the caribou-crazy leader. Unable to hold himself back, he took off after them like a bat out of hell. Well, I thought, go right ahead you bugger and play yourself out. We had plenty of time available, and so I let him go. In a very short distance, the caribou stopped to look at us. I stopped the dogs—they were accustomed to stopping this way so I could shoot. But this time the meat wasn't needed, so I didn't shoot at them. Killing meat and letting it go to waste wasn't our nature, and nearly enough meat to last the winter was cached to the south.

After watching them for a few moments, I swung the team around to pick up the trail again. Unfortunately, when I turned the team around in what was supposed to be the right direction, the leader couldn't pick up the trail. At the time, a ground drift was on—a usual occurrence during the Barrens winter to which we paid little attention. And I was no longer a greenhorn on this, my fourth year on the Barrens, either. But a man never gets too smart to learn, and this proves it.

I never hit my second out-camp that first night. I made a snow house and crawled in for the night, with nothing to eat and no wood for a campfire. With little or no concern about the situation, I'd sort of laughed the whole thing off, figuring we'd hit the main camp the next day. I thought I'd swing around and go right to it.

Next day I swung around all right, but instead of finding the main camp, I hit MacKay Lake about dark, just as a blizzard was bearing down. Mind you, at that time I wasn't aware this was MacKay; I figured that out later after having time to think about it. I'd been traveling too far west all the time.

Across the lake, wood was visible along a sand range. Not wanting to spend a second night without a warming fire, I decided to head for the wood. After reaching the lake, I saw a herd of forty or fifty caribou ahead of me. The ice on the lake was covered with considerable snow, making it a smooth surface for travel. Able to move right along on the lake, the curious herd headed right toward me.

Since this was the second day without food for the dogs and me, one of the caribou had to be shot. I had to make sure of one thing—that the dogs could eat. Otherwise, my goose was cooked. The rifle was out and ready to go when the caribou were in range. I lined up on one and when the rifle cracked, two of them dropped.

I drove up to them quickly because the storm was bearing down on MacKay Lake and I wanted to get to the wood before it hit. I took the hides off and kept them, then loaded the meat on the toboggan and beat it for the wood patch. I arrived just in time, because the weather was really getting dirty.

The very first thing I wanted was a roaring fire. Even though I dug up a rotten birch, I couldn't begin to get a fire started. Either the wind was blowing too hard or my hands were shaking too much, but I couldn't make it go. Having lit match after match, I finally gave up.

I was hungry and had meat, but there was no fire or cooking pan. The only food I could think of eating were the two frozen caribou tongues I'd taken from the kill. I cut these into small chunks and, by God, ate both of them raw. That was the only time in my life when I had to eat raw meat. And, by George, it was two days before I became hungry again. That raw caribou tongue stayed with me like no other food I've ever eaten.

Below this patch of ground spruce was a kind of hollow. After turning the dogs loose, I dug a big hole into the side of a large snow bank. Since there was plenty of meat for the dogs, there was no danger of them straying.

Drowsiness was taking control of me, so I took White Brandy and crawled into the snow hole. I'd selected the big white dog to keep me warm because he was sort of a pet. That hole I'd dug in the side of the ravine was big enough for a bear to roam around in, but with my parka on and the heat from White Brandy, it got so danged hot in there I had to move out.

Next, I dug a tunnel big enough for my body alongside the sleigh and put my tarp into it. I lay down in this crevice and wasn't even slightly cold. And I've never been so sleepy in my life. With the parka hood pulled over my face and my arm under my head, I figured sleep would come and I'd be all right. But I was afraid and couldn't get to sleep. I started to sweat and had to get up. What a helluva night that was!

At the first streak of dawn the next morning, I could see sun dogs to the south and knew it was going to be a doggone dirty day. I had a decision to make, so to the south I swung. Before changing course, I had been traveling southeast.

By holding to the southern course, I knew some big lake would come up and I'd recognize it. Apparently it did, too, because Hughie spotted my trail on Backs Lake as I passed within a mile of the main camp and didn't know the country. The harder I looked for landmarks, the more everything began to look alike.

Making danged sure there was wood every night, I kept going south. Once beyond MacKay Lake there was timber here and there—ground spruce—and I'd head for it. Maybe it wasn't on my course, but that made no difference. I sure didn't want to chance getting stuck out in the open, so I stayed close to the wood.

I kept this up day after day, as sleep kept trying to get control of me. Too bull-headed to give in, I never slept for nine days. Several times I felt myself dozing, but worked out of it in time to fight the feeling off. Kneeling over a fire with my knees on the two caribou skins I'd kept, my nights stayed sleepless.

There was one time just before hitting the last patch of wood near Charlton Bay that I almost gave in. I doggone near dropped off to sleep, and if I had, it would probably have been permanent. My body was warm as toast and crying for sleep. Even though exhausted and feverish for two days, I forced myself to overcome these odds and stay awake. Trying to shake this fever, I walked ahead of the dogs breaking trail.

That fever was a funny thing and probably came about because of drinking snow water. Whenever ice was unavailable, I filled the shovel was snow and held it over the fire to melt. I drank the melted snow from the butt end of the shovel. Unlike ice or lake water, snow water is unsafe for drinking. It doesn't taste the same as good water, and more than one man has gotten sick on it. A man should boil it before drinking, but I never took the time.

The fever was bad and the accompanying headaches terrible, but gradually it eased up. This trip proved what a person is capable of when he has no other choice. It never dawned on me at any time that this might be the end. I was stubborn enough to say I was going to make it, and so I kept going, never allowing complete discouragement to enter my mind. My biggest worry was the weather, because getting holed up somewhere for a day or two would've taken care of me. I prayed that it would be clear enough so I could see to travel. By God, I took that southern course and stuck to it with an unswerving dedication. It was sometimes necessary to stray in order to find the wood patches for a nightly fire, but my senses were never so dull that I couldn't right myself the next day. I could always see the timber four or five miles away on a hill, and always headed there before dark.

The two caribou I had previously killed meant the dogs had plenty to eat for the first few days. But as the meat supply dwindled toward the sixth or seventh day, I had to kill a dog to feed the rest of them. I selected White Brandy and killed him with the ax. My mind was running around in circles and I thought I had to do it. It may seem strange that my pet was the one I chose, but survival depended so much on being practical, and White Brandy was the most expendable animal. I picked him because, after having been put in the harness for the day's run and hitting the trail, he would become lame in one shoulder and had been that way for a year or two. My brother used to drive him and all of us thought it was his collar. We'd tried several different ones and had found nothing wrong, so we'd given up and had been driving him anyway. After being on the trail for a ways, he'd limber up and be all right. But the next morning, it was the same thing all over again.

Since he was the least healthy member of the team, I killed and skinned him, taking out his insides with the intention of eating his liver myself. I quartered the carcass and laid it out to freeze before feeding it to the other dogs. Hot meat was bad for them and they couldn't keep it down. As I was cutting the liver into small chunks—I was going to barbecue it on sticks near the fire—my knife struck something hard like metal. I sliced around it and, to my surprise, found a rusty buckskin needle imbedded in the liver.

This, then, was the apparent reason for the poor bugger's lameness. No doubt he'd swallowed that needle with a chunk of meat at a camp somewhere. It was one of those three-sided things about an inch and a quarter long that we used to sew buckskin garments. It had worked its way through White Brandy's stomach and become lodged in his liver.

The spirit of that white guy had to be tremendous because he had to have hurt fiercely. I'll never forget the way I felt after learning why that poor dog had suffered. A man got close to his dogs, and didn't like to see them in pain. To make matters worse, I would never have had to kill him. But I was lost and out of shells, so I knew a dog would have to be sacrificed so the others could live. I could see no way around it.

That night came a helluva blizzard, with the wind howling loudly and blowing like a twister. I couldn't see or hear a thing. But it eased up by morning, enabling me to hit the trail at sunrise. After traveling four or five miles, I could see ravens circling above something and figured it had to be a fresh wolf kill. This was a chance to get some meat, so I chawed the dogs over to it.

Sure enough, I spotted a dead caribou—a cow with only one quarter eaten away. All the evidence indicated a lone wolf had found her asleep during the previous night's blizzard and had killed her before she could get up. Three or four other cows had remained asleep not over a hundred yards away and never even heard the commotion. The noise of the storm had covered all sounds of the struggle. Anyone doubting whether a lone wolf could kill a caribou should've seen this. And wolves don't always run them down or pick on the sick and weak exclusively.

This time the wolf's bad deed was a blessing, as the dogs and I needed the meat. Finding that dead caribou had a sobering effect on me because it was only the night before that I had elected to kill White Brandy. This made me very blue indeed. If only I'd waited until morning, the big white Husky would've been with me still. But I wasn't in my right mind, and my thinking was clouded.

The dog team and I lived on the two caribou I'd killed plus the wolf kill for nine days and nights. I never had a bite of anything except plain caribou meat the whole time. I placed the meat on twigs stuck in the snow, and barbecued it by the campfire. Because of barbecuing that meat so close to the fire, my parka sleeves later had to be slit with a knife by the Mounties at Fort Reliance to get it off. The caribou-skin garment had cooked right along with the meat, causing it to shrink and shrivel until it fit like a glove. At the time, I was unaware that this was happening.

Since the dogs had food and were better rested than I, they were in pretty good shape. When getting to the hilly country where the deeper snow was, my leader had to break trail by jumping up and tamping the snow down. I had become too weak to go ahead of them, even though I had trail shoes. My luck at having found meat and my bull-headedness were what kept us going.

By staying on a southern course, McLeod Bay, which is a full ninety miles long, would be impossible to miss. Since I wanted to hit it toward the eastern end, I adjusted somewhat to a southeastern course. This wasn't an impossible task, because I did have a compass. And no matter how foggy my mind became, I believed that instrument faithfully.

The course was an accurate one because I came to Walmsley Lake and knew immediately where I was. The lake was familiar because I'd cut across it on my first year in the Barrens on the way back to Fort Reliance in the spring. Hughie, Magrum, and I had come back that way from Aylmer Lake. I was a very relieved man to be in familiar country. I was convinced of my location after I spotted the lone tree where the lake flows into the Barnston River by open water.

I kept going in a southeastern direction and came out where I wanted: the Hoarfrost River. This I was sure about, too. I saw beaver and otter sign, and across the way was Sentinel Point, some twenty-seven miles distant. Still, I couldn't fully believe myself.

Even after coming to McLeod Bay, I kept looking at Sentinel Point across the way. It just didn't look right to me and that filled me with doubt. But running through my mind was the thought that it just had to be the place because it's a big hill, clear-cut and long. It was still about ten miles from where I struck McLeod Bay.

Next, I tried to figure out this point that looked so hazy. Even though it was twenty miles away, I knew it must be Maufelly Point. I looked for a further landmark to confirm, and saw the three drops on the horizon. But, still, it didn't look right. While trying to set all these landmarks straight in my mind, a lone wolf trotted along beside me, brave as you please. For some reason, he held no fear. Perhaps he sensed that I was out of ammunition and couldn't shoot him.

Finally, far enough down the bay to see the three ridges plainly, I knew Fort Reliance couldn't be far away. Too many things in the terrain checked out to make my location a coincidence. Besides, there are no two places the same. Still, I'd been lost for so long and wasn't absolutely positive until I was rounding the point and saw the police barracks before me. There aren't words to explain how deeply relieved I was as the team took me toward the comfort of those buildings.

After landing at the police barracks, a surprised Mountie asked me what in the world I was doing back from the Barrens so early. He shook his head in disbelief when I told him about getting turned around at my last out-camp and that, unable to find any of my camps, I'd struck out for

Fort Reliance. He didn't doubt my word when I told him I hadn't slept for nine nights. I suppose I must've looked it.

I remember stopping him when he started out to get my bed from the sleigh. I told him he'd find no bed or frying pan, only a snow shovel. Since I was so weak, I had become somewhat incoherent. It was just through sheer bull-headedness that I'd kept going. If I had gone to sleep, I would've dropped just like that. Deep inside, I had been afraid to sleep, so, by the Jesus, had stayed awake for the whole trip.

After cutting the parka off, the officers fed me lightly with some eggs, toast, and milk. Afterwards I became so weak and relaxed that I staggered drunkenly. They gave me one of their eiderdowns, and I was placed on their couch to sleep. Funny, the whole miserable ordeal had come to an end and I was absolutely exhausted, but I still found it impossible to sleep immediately. I lay there in a daze for a long time before finally relaxing enough to go under.

For two days, my bed was a mess when I awoke in the morning with pieces of skin—that looked very much like the shed skin of a snake—everywhere. Since I was very groggy, it never dawned on me what it was. I kept picking it up and throwing it aside without knowing or caring. Since I felt no pain, I wasn't too concerned. But when my senses finally started coming back, I noticed this stuff and asked one of the Mounties what it was. He told me to look at it closely because it was my skin, and I had been peeling like a snake. I investigated and saw that, sure enough, it was coming from all over my body. I blamed it on the excessive sweating during those two days I had the fever.

It was several days before I could move and think normally. It was too late to try going back to the lines and too early to head south, so I waited around Fort Reliance until Hughie came out in the spring. Even though I had come up missing, he had remained in the Barrens until the end of the trapping season. That's all he could do. Since he had a lot of fur to fix, his trip out was delayed. He and the Magrums came out by way of the Barnston River across from Back's Lake.

Instead of cutting across McCleod Bay toward the gap and heading for Snowdrift, they had cached their fur along the way and come to Fort Reliance to report me missing. This was the first week in May, almost six weeks after I'd disappeared from the camp. Traveling on the ice by dog team was still very safe and easy.

When Hughie got in and saw me, he didn't know what to say. After staring for a while, all he could think of was, "What the hell are you doing here?" We sat a while and I told him the whole story and then, in turn, received his version from the other end.

When I hadn't shown up on time back at the main camp, he had gone out over my line to see what had happened. Magrum, who was visiting at the time, had returned to his camp while Hughie went looking for me. After arriving at my second camp and seeing my eiderdown, the tied dog, and no sign of me, he knew there had been trouble, figuring I'd been away five or six days. This was on Saturday and I'd gotten lost the previous Monday.

He considered the possibilities and concluded I'd drowned in Icy River. I wasn't even aware that the river was open in the winter, because all I'd ever seen was overflow ice. Anyway, Hughie cried. He told me so himself. I suppose that would be a helluva shock to imagine that fate for someone—and your own brother, at that.

So, Hughie made a trip down to the river, but of course found no sign of me. There wasn't even a trail over my line, but the lead dog took Hughie over it because I'd driven him there before. The bugger stopped at all my traps and Hughie picked up some fur from them.

After his trip to Icy River, he returned to my out camp and picked up the spare dog. He found the caribou I'd killed, the ones by the camp. That dog was still tied with the rope even though meat was only a short ways away. For some reason, he'd never chewed his way clear, which meant he was one hungry hound, indeed. He had stayed right there for a whole week with nothing to eat.

Right away, Hughie went to Magrum's camp, hoping I might have gone there. But with no sign of me, they struck out and made a big circle. At that time, they found no toboggan trail of any kind, but saw my trail on Backs Lake a long time after that. They realized that I was headed southeast and was on foot, but the trail was very light and, finally, they lost it altogether. Hughie said he doubted if I passed more than a mile from the main camp.

Although I never recognized the lake, I recall at the time that I thought it had to be Backs Lake because I'd killed some caribou near those islands and set some traps by them. I had looked, but hadn't been able to find those danged caribou. Because I couldn't locate them, I couldn't confirm that I'd arrived at the lake, although I felt it was right. Finding the meat and traps would've enabled me to be in a nice warm bed that night. But everything was so much alike out there that I wasn't sure of myself.

That was the toughest trip of my life and I don't like to talk about it. I almost lost my life, but was too stubborn to quit. At that time, I was pretty tough and it was a damned good thing I was. Even though that was forty years ago, I can still feel myself kneeling sleepily on the caribou hides over those little fires. I remember how it felt when my arms started to go numb from fatigue. I'd catch myself and fight it off. I found out the hard way what a darned tough creature man can be when he has to be.

I think a man can condition himself to stay awake nearly all the time, but he can't let his mind run away with him. In my case, my thoughts never strayed from my purpose, and determination pushed me to accomplish my mission. Leaving before sun-up after a sleepless night in an unfamiliar country doesn't give a man much to look forward to, but I forced myself to keep track of the sun. It rises and sets toward the south up here in the winter, and so that fact kept me pretty well on my course.

At that time, I was still getting along pretty well with my family. Thinking about my two little girls in Vancouver sure helped. I discovered that a man's mind reaches out for the things he cherishes when the going gets tough—and, believe me, that trip was a tough one.

After getting in a tough situation, it's easy to look back across those small decisions that could have changed it all. There was no reason for becoming lost, but I became careless. Figuring I was saving time, I didn't bother to go back over my trail. All I had to do was turn the team around and walk back over my trail to make the leader understand I wanted to go back the way I came.

I could've forced the leader to take the trail but, no, I yelled "chaw." Well, he turned all right, but in a direction that took us away from the trail all the time. And the first guldanged thing I knew, I was lost, and never able to get untangled.

There was a ground drift on and the trail would've soon been lost, but the dog didn't follow it. Never would I make that same mistake again. Backtracking is the slow way, but out in the Barrens, no way but the sure way should ever be taken. Caution is the secret to success and comfort, but I threw it out the window just to save some time. And time was the one commodity I had the most of.

The next fall before going back to the Barrens, I put traces—runners—on the toboggan with heavy bolts, to ensure they'd stay affixed. They leave a deep track and, if they'd been on my toboggan during this misadventure, my trail probably would've been visible at some points. It also makes the toboggan slicker, and dogs can pull it much easier. If I remember correctly, I never went back to the trap lines again without a set of them.

CHAPTER VI

With 1933 and all its miseries in the background, I was ready to seek new country, and looked toward the northeast for the change. The name D'Aoust had become common to the Barren Land after Hughie and Phil first came out in 1929 and I followed a year later. Hughie and I had been together for those first few years, but the time had come for a change and, with Phil, I headed to the Whitefish Lake area east of Fort Reliance. I wasn't aware of it at the time, but the area northeast of Fort Reliance was to become my trapping home. The rivers Hanbury and Thelon, and the lakes Whitefish and Beaverhill, would call me back year after year, and I would stay to work this same general area for the rest of my trapping days.

Predictably, my family associations became less and less of a factor in my life, and were reaching the inevitable fate for which they were destined from the outset. My wife was a fine stenographer who worked for the government and moved around the country frequently. It had become very difficult for me to keep track of her. My life was one of remoteness and communication with me was difficult and slow. Her life, meanwhile, was filled with travel within the populace of the South. We were alike in one respect, however: both of us needed to be on the move. The difference was in the places we liked to do it. We simply drifted apart because of the factors separating us. But primarily because of our religious beliefs, divorce was never to come between us.

As this, my biggest tie with the outside world, was slowly severed, I began to adopt the North as not only a place to trap, but as my year-round home, as well. I still made periodic visits to the outside world, but they became less meaningful as the years passed on.

One particular place, Charlton Bay, at the extreme northeastern end of Great Slave Lake, became a place of special meaning for me. This bay was, and still is, the most beautiful place I've ever viewed. All I've ever had to do is stand by my window and look around and across it. There can't be any more pure or unspoiled magnificence in all of God's earth. Every time I've looked upon it, I've seen and felt things like I have in no other place.

This is the first bay to freeze in the fall and the last to thaw in the spring. A man can look over it in the summer and all he sees are green trees. It's like an oasis, and yet it's situated only forty miles from the Barren Land. It doesn't look like there are any birch at all, but they show up after the first frost—and they're beautiful, too.

The very first time I cast my eyes upon this bay, I fell in love with it and felt safe as it surrounded me. Afterwards, whenever I would describe the geography of the North to someone on the outside, I would talk about this bay. Its waters sparkle and glisten because they're so danged pure and clean. The freshness of the water is guaranteed by the thrust of the Lockhart River, as it pours from God's country beyond. There's just no doggone way that river can be polluted. And the place teems with trout.

Once, I had to fight a verbal battle for my bay with the Mounties. That was probably twenty years ago when they were stationed at Fort Reliance and trapping was still a way of life. I had to get on them because they used to throw all their caribou and dog hair, as well as dog manure, onto the ice to get rid of it in the winter. They figured it would float out with the ice in the spring. But the waste neither sank nor drifted away. Instead, it collected back in the little bay by my cabin. Hell, this was the source of my drinking water and my place of perfection on earth. I couldn't sit around silently while the police dirtied it up. Once I got worked up enough, I went to them and relayed my feelings in direct terms. By golly, they appreciated the fact that I was concerned and stopped doing it. I guess they just didn't realize what they were doing.

In the old days, as many as seventeen or eighteen of us old trappers were going through this bay on our way to the Barren Land every fall. In the beginning, we went the difficult and slow way—over the portages; in later years after the float plane and bush pilot had become a way of life, we'd charter a plane and fly out to our camps.

All the trappers would wait around until time came to fly out. We were cozy and comfortable all camping together in what was a recurring city of tents. This was affectionately referred to as "Trapperville", and everyone was happy to be a part of it.

Probably as a precaution more than anything else, the Royal Canadian Mounted Police had built a post at Fort Reliance when the white trappers first came into the country. There never was any trouble, but they built a barracks and stationed troops out there, anyway. Everyone seemed to get along in this community with all the tents and equipment around the barracks. The Mounties stayed until the white trappers pulled out, then they left, also. Fort Reliance got its name because it was a jumping-off place for the Barrens beyond for a group of explorers in the early days.

Trappers would start drifting in from the South around the fifteenth of August. By the first week in September, the time would have come for getting to the trapping camps in the Barrens. In the early days there weren't many planes, and it took a while to get everyone out. Just two or three flights a day were flown—and then, only if the weather was fit. Since all of us were very anxious to get out there, a fair way was worked out for using the plane charters: names were drawn from a hat to determine the order of flights.

By the time everything was in order and we'd reached our camps, a good frost or two had spread over the Barrens. The pilot would put us down as close to our main camp as he could, choosing the closest big lake safe enough to land on. In most cases, mine was well away from a lake, and so my gear had to be hauled overland for a distance.

By September, the air had cooled considerably. That took care of most of the bugs, and made the Barren Land a pleasant world to be in. Over all the flats and ridges on the tundra at this point in the year, the vegetation would change rapidly into its fall colors. Even though this wonderful scene would be visible for only a short time, it was a beautiful, beautiful sight to behold. The colors were bright and vivid with many contrasting patterns and hues. Because the air was crystal clear most of the time, my lungs could never seem to hold enough of it. Being out there at that time of year is refreshingly healthy and always makes me thankful to be alive.

The low bush cranberries covered the low hills like a pitted, thick carpet of glistening red on green. No man with a sense for things beautiful in nature could ever forget how this looked. And those cranberries! As a special treat, there's nothing quite like a cranberry pie. You can't beat it.

Also native to the Barrens were other berries, yellow ones that grew on damp flats whose name I never knew. Although not overly sweet to the taste, they were good eating, and I always felt they were good for my system, too. I filled up on them many times. They'd stick to a man's ribs, but I learned never to boil them with sugar; they'll work on a man and keep him going outside all night long.

Further study of the land at this time revealed that other places were still green, especially those supporting mosses and lichens. Some of the low spots were more colorful because the frost seemed to have its effect there first. Yes, everything was preparing for winter, and it wouldn't be long because autumn in the Barrens was a very short season, indeed. I've seen several years when fall didn't last longer than three or four days. Summer had always just seemed to quit when Old Man Winter took over with a grip of permanence. It could start snowing any time after the first frost. And when it does snow out there, it stays, not melting away like it does in the South.

I liked to walk and would inevitably find my way to the top of a ridge. Looking around in all directions, the vastness of the country was awesome. It's wide and big and seemingly has no end. The first time I saw this, it was very frightening. It let me know right away that I'd better respect it. That big country beyond the trees was no place for a greenhorn, especially during the winter months. No sir, it was no city park.

If fortune was smiling and we hit the right place at the right time, caribou could be seen everywhere a man looked. With the first frosts out on the Arctic tundra, the small, scattered herds would start gathering into bigger ones. Pretty soon, a big storm would spread across the tundra to the north, and send the big herds southward on their yearly migration to the trees. It was a magnificent sight to witness: caribou, spread out as far as the eye could see, moving like a wave toward their wintering grounds in the south. On and on, they would continue in an endless line.

As a rule, September nights were very cool, but the daytime temperature usually improved sufficiently giving the air enough warmth for all the comfort a man needed. There was some rain in the Barrens during September, as this seemed to be the rainy month. A few times, my charter landed in the rain. After pitching my tent on the shore with my equipment nearby, I would just sit tight and wait until better weather made traveling easier.

The hardware caused little worry and was cached right near the landing, but the first concern was for all the perishable foods which were vital for a comfortable existence. When the weather became fit for traveling, I would begin moving the perishables. Everything had to be moved to the cabin which served as the main camp.

Even without snow on the ground, the dogs were put to work moving the supplies. Out there, the vegetation was slick, and the dogs could poke right along with a load. They had little trouble moving the sleigh loaded with 250 pounds over the tundra. They were put to work because that's why I took them out with me; besides, they enjoyed working and were bred for it.

Food and equipment for living and trapping seven months made for a sizable pile that couldn't be hauled in one trip. The whole process took some time to complete, and certain bagged foods had to be protected from mice while I was away between trips. Other animals were a nuisance, but those guldanged mice seemed to be at their worst in the fall.

By making a stage, the food could be kept above these pests. After building a rectangular frame with a long sturdy pole in each corner, I was ready for them. I stretched a tarp over the frame to form a platform and then placed the food on the stage, well beyond reach of the mice. To keep them from climbing the legs of the stage, I nailed a tin can around each about a foot above the ground. The mice could climb to the can, but not any further because they couldn't get their claws into the tin.

Enough trips—a couple each day—were made between the landing and cabin to get everything where it belonged and, finally, the work was done. And danged hard work it was, too. Mind you, the work didn't bother me a bit; it's God's country out there, and a man feels like working.

If it was too soon for the caribou migration, I would wait for that big freeze on the muskeg to send them south, hoping they wouldn't bypass me. This instinct to migrate sent them to the trees to rut. Each year was a gamble, with us set up in early September between the caribou to the north and the wintering grounds to the south. If the herds missed us, it was either a long, hungry winter or a tough, hungry trip back south by the first of November. We simply could not get by without the meat they provided. A dog team and a man needed lots of meat over a seven-month period. Without caribou, Barren Land trapping was impossible. We needed between seventy-five and eighty to last the season. While killing that many animals might no doubt seem like a lot to most people, there was nothing else to eat out there.

Besides, I figured a pack of five wolves would bring down and kill that many in five months, and I danged sure took more than five wolf skins every year. According to my calculations, I stayed ahead of the caribou I killed in the amount of wolves I took. I was way ahead of the game, and no wolf lover better argue with me, either. I learned something about the two animals in the nearly forty years I lived with them.

By the time the caribou arrived, all the flies and bugs were gone, and hunting had begun in earnest. Soon after killing them, I cached the caribou—mostly in ground caches—until I needed the meat. Spoiling was no problem once the flies were gone, and the meat kept quite well all winter. The process meant that, as soon as I killed a caribou, I'd cut his throat and gut him. I'd pile five or six up together and cover them with a layer of brush. Then, to keep animals from getting to it, I'd put rocks over the whole thing. After the meat became frozen, the layer of brush would keep the rocks from freezing to it. When the meat was needed, it was easy to get. I just had to pull the rocks away and there it was, fresh as the day it was killed.

With plenty to do before trapping began, few trappers were ever bored. Most of us had years when we were completely alone. For me, it was tough for only the first couple of days. Then the feeling of solitude would pass and, by the Jesus, I would become elated. I don't think I have ever experienced such happiness as I did during those times I spent alone.

I might go for a period of seven or eight months without seeing a plane or even a toboggan trail other than my own.

But although I was alone, I was never lonesome.

Mind you, all of these happy times weren't without their moments of doubt. One year, everything went along well, but I could never get enough caribou. A feeling of deep concern came over me because it was getting late in the season and I'd only killed thirty-five head—not nearly enough meat for the winter. I had to start hunting hard or my season would be a short one.

It was September twentieth, well past the time when the meat should've been cached and forgotten about, and I was hunting well beyond my camp. All that morning I'd hunted hard without seeing anything. Not even a hare! And then early in the afternoon while looking along a ridge, I saw six dry cows headed directly for me. Waiting motionlessly until they were well within range, I opened up on them. I was shooting my .270 Winchester, and when that rifle lined up, something happened for sure. Five of them dropped before they could get away.

The meat was desperately needed and it had to be safely cached, but I searched everywhere around the spot of the kill and couldn't locate anything from which to make a cache. Brush and rocks were simply nonexistent. I had no idea what I was going to do. Just the same, I gutted the carcasses and put them in a pile. Even though it was four or five miles to my camp, I headed there to try to find something to protect the meat.

Once there, I looked around, but couldn't find anything suitable. It was then that a crazy notion entered my mind. Carrying four short poles, some wire, and four empty jam cans, I went back to the dead caribou. Once there, I drove the four poles into the ground, one at each corner of the meat. I attached a single strand of wire to the poles about a foot off the snow and halfway between each pole, then hung a jam can to dangle freely above the snow. I left that pile of meat exposed with only the single wire and the four cans to protect it. I didn't know if it would work, but was at a loss for other ideas.

Early in February, my meat supply had dwindled, so I returned to this cache. Not only was the meat intact, but no animal had gotten inside the flimsy fence. Wolves and foxes would go up to it, and then turn away from the bobbing cans. Both old and fresh tracks were everywhere around the cache, but every animal had lacked the nerve to go inside. There was even a set of wolverine tracks around it, but that little devil had also shied away. When I told another trapper, Matt Murphy, about this innovative cache protection later, he wouldn't believe it.

My brother, Phil, never hesitated to make a cache out in the open, either. When materials were unavailable for a ground cache, he'd anchor two poles with a cross-arm between them. Quartering the caribou, he'd hang the meat on the cross-arm high off the ground so the wolves wouldn't get it. Even though this was right out in the open and could be seen for miles, Phil got by with it.

Another year, still early in September and during the hunting season, Delphine (who was my second wife, and whom I will introduce later) and I had struck out for a ridge north of our cabin at Beaverhill Lake. Before we got to the top of the ridge, two cow caribou had crossed in front of us. Even though they were at long range, one of them dropped at the crack of the .270. I wasn't aware of it at the time, but these two cows were the forerunners of the big herd to come.

This was early morning and, by eleven, the cow was dressed and safely enclosed in a cache, and we had gone on toward the ridge. Just as we approached the peak of the ridge, a herd of possibly a hundred came over the crest right at us. Those marvelous animals were traveling the straightest bloody line, characteristic of them during migration. Because their curiosity was stronger than their fear of man, we were able to walk right up to them. With meat all around, I went to work on them with my rifle, dropping seven bulls, and were they ever in great shape. In quick order, we gathered rocks and brush and the meat was cached.

Taking time out to fix some lunch and boil tea, we felt good, like everyone always did out on the trap lines when they knew for sure that the caribou weren't going to be missed. When our lunch was completed, we continued on up to the top of the ridge. Looking casually around from the summit, I became alert immediately and could hardly believe my eyes. Most people have never witnessed such a spectacle. The whole country was a moving mass of caribou. Looking as far to the north and east as I could see, the tundra was a solid wave of moving animals.

Without wasting any time, I waded into them, intending to get the meat supply cached and to eliminate any worry about shortages. I'd shoot for a while and then, with Delphine's help, gut and cache the dead caribou. No sooner would the work be done than another bunch would come along and we'd repeat the performance.

Delphine later kept track of the shooting and, from a box of twenty cartridges, I knocked down nineteen caribou. Around the same ridge in just two days, I took forty-five caribou, nearly all were bulls in prime shape. These caribou were the fattest I'd ever seen in all my days on the Barrens. The herd was headed south to winter below Granite Falls at the tree line.

This is the way September would go for the Barren Land trapper who spent the month hustling to get his meat. None of the years were alike, in either the appearance of the land or the things we did upon it. Sometimes, the caribou were scarce because they passed somewhere else. Sometimes, winter came a little sooner. But generally, this was a happy month that concluded with my anxiousness to get out to my camp, where I would feel nothing but happiness. Trapping was a blood thing with me that never left my system.

By the time October rolled around, the meat supply was hopefully already secured. The weather was still pretty nice early in the month, but the temperature would steadily drop until around the twentieth. By then, with snow having been on the ground for a while and accumulating rapidly as the days sped by, little doubt remained that winter was close at hand.

Taking advantage of the lull in activities, this was the month for working on the wood supply. The main camp needed a large pile of wood for winter cooking and warmth. With the dog team and toboggan, I'd head for the patches of wood scattered here and there, and cut and haul it until I was sure I'd gathered enough for the season. This was slow work, as the wood came from the small trees and ground spruce that grew on the south side of sand ridges running in a general direction of east and west. These small Arctic spruce grew only to a height of three or four feet.

What few trees found out in the Barrens were tiny because they didn't have a chance to grow. The climate was much too cold, and strong winds and blowing sand created further problems to tree growth and survival. The few trees that made it—and the bigger trees were well over a hundred years old—were only three or four inches in diameter with a vast system of roots that ran all over the place. After the wind blew the soil away and exposed the roots, the tree usually died. The dead trees and roots made excellent fire wood. That was real cooking wood.

Since the wood patches were scattered and generally not close to the camp, most of October was required to get this job completed. The work was enjoyable and the atmosphere for doing it ideal. I felt like working with a cariole of firewood as defiance to the impending winter all the incentive I needed to complete this job successfully. Even the dogs, apparently eager to get themselves in shape for the long runs over the lines, seemed to enjoy pulling the loads of wood.

The trapping season didn't get underway until after November fifteenth, about the time the fur became prime. Around the first of the month, I'd string the traps out and attach them to the toggles, but I didn't set them. I'd also run the fish nets at the same time. Fish were a very important change of diet which suited me fine. I could eat them once a day all year. Dogs were crazy for fish, too.

I dried and stored some fish on a stage near the main camp, but I never seemed to have enough. There was so little time for fishing, and it took so many to make a difference. I netted whitefish until it became too cold, and they went to deeper water. Trout then came to the shallows, and several twenty-five - and thirty-pounders, along with an occasional larger one, would find the nets.

Although the traps weren't ready to catch fur, they were attached to the toggles and the sets were made ready for the time they'd be needed. When the season did come in, all I had to do was run over the line and start the traps working. That's the day I waited for because, from then on, it was all fun. The thrill of running traps remained as great on the day I ran the last one as the day I ran the first one.

After what seemed a slow, long wait, the day came for going over the line and setting the traps. This was probably the most important trip, because care had to be taken so the trap would spring when triggered. A frozen trap wasn't a working trap, so I danged sure set them so they wouldn't freeze in. If a man made a bad or careless set, the trap might just as well have been left on the sleigh.

Around the eighteenth of the month, I'd reach the half-day line. By then, the white fur of the little Arctic fox was prime, a pure snowy white. (Before then, the fur had a blue undercoat and wasn't worth anything.) The half-day line became longer as the daylight hours began to stretch out in the spring, so I would leave a load of extra traps and equipment at the end of it. I would gradually extend the line as light permitted.

By then, I'd been over the entire line to make sure all the sets were good ones with traps that would work, regardless of the type of weather. The main line stretched far enough to take advantage of the widely scattered wood patches. While trapping at Beaverhill Lake, that was twenty-two miles. By the time I had everything exactly as I wanted it, the first of December had arrived, and fur was beginning to come in.

The number of traps used varied according to the ambition of each individual. One man tried to make me believe he had 400 traps actively working for him all winter. By himself, too. He said he did it all winter. Since I knew the man well, I didn't openly dispute his word but, secretly, I thought it was a bunch of bull. I had come to know the trapping game quite well, and had made some real fine catches, but the most traps I ever took with me was 150. And I had to struggle hard to keep up with all of them. If a man knows what he's doing, and makes logical sets and he keeps them working all the time, 150 will give him all he can handle.

When all my traps were working, I was at the height of my glory with four months of enjoyment laid out before me. Fur was coming in and I was working the job I knew best. Trapping was fun, not work. I studied the weather in very careful detail and, when it looked bad, I stayed in camp. But the next time I saw a fox dangling in a trap would make up for all the bad times.

Sometimes, a blizzard would snow me in. But there was always something to do in a camp. Wolves and foxes had to be skinned and the pelts had to be dried. Sometimes, I'd simply catch up on some sleep. And baking a bannock, or trapper bread, was always a pleasant chore.

INTO A CUP PLACE:

   2 cups white flour

   1 cup rolled oats, bran or whole wheat

   A dash of salt

   2 tablespoons baking powder

MIX WELL.

ADD:

   2 handfuls brown sugar

   ¼ pound of lard or caribou fat

Mix everything until it's blended well, then add milk (better if soured) until the batter is crumbly. Avoid mixing too long because it shouldn't be runny. Leave it fairly dry. You want it so it can be handled with your hands.

Grease skillet. Use a heavy aluminum, not a cast iron, skillet. Spread the mixture in the skillet, punching down until level. Push away from the skillet sides with a fork, so it can be taken out and turned over. Put on stove and keep fire low. Use a grill or a couple of rocks to keep it off the stove so the bottom won't burn. When you think the bottom has had all it can stand, take it out of the skillet and turn it over. Cook until done. Makes about an eight-inch square skillet full. Nuts or raisins can be added to the mixture. Serve with butter, jam, or honey. That's trapper bread and it'll stick to your ribs.

Also during the storms, the dogs had to eat. I had to go out to a cache and get caribou for them. That was something I never wanted to do, but had to. A man's survival was directly dependent upon the health of his dogs.

Even though the winter months were long and mostly dark, with some danged tough spots, time passed right along. In no time, the day would come to go south with the catch. There was no flying back for me. I drove the dogs out with my catch on the sleigh.

CHAPTER VII

Since Samuel Hearne first named and described the Barren Lands in the 1770s, they have probably changed very little. Physically, the area has remained the same country he walked though in search of the copper mines on the coast of Coronation Gulf of the Arctic Ocean. Much of the tundra I knew was the same as this famous Arctic explorer had known. Towns and settlements were still few and far between and identical extremes of nature existed. Progression had done little to change the geography that the natural evolution of time had so carefully plotted for the Barrens.

A few of the animals that were common in his time have either grown scarce or ceased to grace the earth with their presence. In most cases, man has been responsible for this change. Other animals are nearly as numerous today as ever.

The clean Arctic air is as wholesome as can be found in the cleanest corners of the earth. The water in the cold rivers and clear lakes is as pure as nature can provide. Casting one's gaze to all sides offers a panorama in which the unaltered work of God remains strong, even in spite of man's ability to subdue it.

Although the Barren Land I know is very small when viewing the great vastness of the overall country beyond the tree line, it exemplifies well the cultural change that has come about since the white man first saw it. Much of this change has taken place in the nearly fifty years I have known it. I tend to find it impossible to understand and accept the altered lifestyle of the natives.

When I first came into the country, the natives were a proud and tough lot. They went out in the winter to trap and live off the land, and they did this well, because this was their natural life. Some of the natives trapped in the timber; others went on out to the edge of the tree line; the most hardy went beyond into the Barrens, living there summer and winter, and only coming in to sell their furs and get supplies. And they were happy.

In 1920, the natives I met were still quite primitive. Fort Fitzgerald and Fort Smith were flourishing posts and very important to them as places of trade. The relationship between the traders and Indians was a strong one from which both benefitted.

I got acquainted with the Chipewyans and what were called the Caribou Eaters who came from the east over around the Dog and Taltson Rivers. They lived out there year round, only coming to the posts in the spring. They'd remain for a short time, and then back to the wilderness they'd go.

For years, this is the way they lived, but they've forgotten this way of life recently and have settled in the towns. I made some great friends among these folks, who were excellent trappers—much more so than they are today. Of course, they never got relief or script at that time—not even old age pension.

Today, the Chipewyans can go to schools and get a good education. They can learn a trade with the help of the government. Many take this route and do very well, but others take another, and rely on all the free money available.

In the old days, the trap line was the only means of revenue for the natives. They fished and hunted caribou for food. If they had caribou, they had everything, because it was easy to gather fur. The country wasn't burned out like it is now thanks to all the old forest fires caused by campfires which weren't put out. Some good marten and beaver country was ruined by these fires. The natives trapped to get the extra items they wanted, including canoes, boats, traps, clothing, and other equipment.

There are some very, very fine fellows among the natives. If they're just given a chance, they'll improve themselves. Pride is a deep thing among men, and they still have as much as any men anywhere.

One big problem is the rapid increase in the number of beer halls in the North over the last few years. Some of the best men are unruly and unreasonable when liquored up. Having seen the effect of the combination for a lot of years, I can say that liquor and Indian blood just don't mix. I don't want them around me when they're boozing it up. In all fairness, there's no difference between them and a white man when they're drunk, but a white man doesn't seem to react with as much hatred and violence.

Since the ways of the Indian have changed to a more socially structured style, they no longer go through Fort Reliance to the trap lines. Once in a while, a native around here will say he's trapping, but he won't be running more than a dozen traps. Even then, he probably won't go out and check them. That's not trapping.

The natives quit the lines at about the same time the white trappers pulled out. These white trappers got too old and nobody came to replace them. As such, there is fur out there dying of old age and neither natives nor white trappers will go after it. With the high price of long fur, there's no reason for anyone to be on relief. There are 500 miles of Barrens out there and I doubt if there's anyone on it.

Back in the early thirties, a large group of natives once camped on Artillery Lake on the west side around Timber Bay. There were also a couple or three families on Whitefish Lake and on out into the Barrens. It was the same scene out of Stoney Rapids. But when the white man quit, they all pulled out of there. It's all wild country now. In fact, it's probably wilder now than it ever was: there's nobody out there and lots of fur.

The native of the early days obviously no longer exists and I think it's a shame. He no longer travels with the caribou herds and lives the happy nomadic life he once did. Though the land is still sparsely populated with plenty of room for all, the native has elected to primarily live a life made easy for him by the ways of civilization. He lives in the small towns, close to the links of so-called progress. This is difficult for an old freedom-loving trapper like me to understand. I've always endorsed a belief that a man should never go against his nature. And I believe the native is now living in an environment that's directly opposite to what is natural for him. It grabs at me to see a way of life coming to an end.

I know and have felt this native freedom by living well away from civilization in this wonderful country. As I came to love and respect this land, I could speak of it in a positive way.

You know, I can't figure why they don't at least go out into the wilderness in the summer. I once stayed out in the Barrens for twelve months without coming in and that summer was the most enjoyable I've ever spent. I had the freedom of spirit all men dream about. And the experience gave me the opportunity to understand the way of life the freedom-loving natives of yesterday enjoyed.

With a man named Red Noyse, I spent that summer on the Thelon River northeast of Fort Reliance. This river may be the most beautiful in the North, an opinion I know other men share. This beauty plus the unending solitude made for a completely worry-free year.

The Thelon River is characterized by a thin line of timber along its shores and water so clean it sparkles. Delphine told me "Thelon" comes from a Chipewyan word meaning "last woods river." I've spoken the word again and again, and always find it to have a poetic ring. The Chipewyan sang a song that celebrated the river. It was kind of a chant and it rang well in a man's ears. Delphine was Chipewyan and could sing it very well.

I remember well my first trip down the Thelon with Hughie. We were headed for our lines with all our gear and supplies in an eighteen-foot canoe. It was a very big load, and we were trying to make some fast time with it. We hadn't gone very far when the first set of rapids was before us. Not taking time to look at them first, we blasted straight ahead without hesitation. That was a mistake, because the buggers extended for a good quarter mile and were pretty rough. I was in the front and it looked like we were going to hit every rock along the way. Even though the water was quite clear and very deep, those rocks below looked close enough to hang us up and cause a spill into their cold teeth.

Arriving at the next set of rapids, I talked him into pulling to shore so I could walk on ahead to look things over. When I returned to the canoe, I told Hughie it would be a rough bugger to get through. He wasn't scared of anything, and insisted it would be fine. I agreed with him, but still had my doubts. Figuring to play it safe, the rope which was tied to the bow of the canoe was coiled by my feet. I told him I was ready to jump out as soon as he yelled. Sure enough, no sooner had we started through and he shouted. Over the side I went like a jack out of a box and bloody well held onto that rope tightly, too.

That water was up to my chest and damn was it cold and swift. From that point on, we portaged around all rough water, taking no chances on losing our rifles and outfit to the charging Thelon. There wasn't that much rough water until Granite Falls, but still we played it safe.

It was in April that Red Noyse and I left Snowdrift for the Thelon River with our two dog teams plus Phil's dogs—seventeen in all. In a couple of days, we made it to Fort Reliance, and lay over there for two or three days because of a snowstorm. On the twentieth, we left there for the Thelon.

The previous winter, Phil had spot-trapped out there, and so had Red. The area and route were familiar to us and, we were able to reach Whitefish Lake in short order. That first night, we camped thirty miles beyond Fort Reliance on the edge of the Barrens. Caribou were coming out of the trees heading for the tundra. It didn't matter where we looked, there were caribou.

We planned to make it to a well-known draw the next day. This was on a peninsula and the next patch of wood on our route. On a big range to the south about half a day away from the draw, we saw a herd of seventy-five to a hundred caribou. I wanted to stop and kill one, but Red insisted there'd be others at the draw where we were to camp. Since we had to move right along to make the camp, I agreed. But after arriving and pitching the tent without sight of any more caribou, we became resigned to the fact there weren't any more. The dogs went hungry that night. To make things worse, there had been no meat at Fort Reliance and the dogs hadn't been filled up for several days. An April snowstorm added its dirtiness to our gloom. The store-bought food wasn't doing much for the dogs' stamina as our loads were quite big.

We headed for Sandy Lake the next morning, about eighteen miles away, and sought a cabin on the north shore. Red was worried about his hungry dogs and said that this would be his last day unless we found caribou. I told him to suit himself, and that I was going on regardless. I'd made up my mind to summer on the Thelon River to work on that cabin and trap there the next winter.

Soon after our little discussion, a small herd of caribou came running by us. They were almost out of range on one of the Sandy Lakes, but after several shots, we somehow dropped one. That saved the day for Red, and his mind was immediately changed. That night, we made the cabin and had quite an enjoyable evening. The following day, we really ran into the caribou as they crossed northward between Sandy Lake and Whitefish Lake by the hundreds. Even though we still had twenty-two miles before reaching the Whitefish Lake cabin, one of our old camps, we stopped to make sure we got meat. The one caribou we'd bagged the day before wouldn't last very long.

Traveling by this time was becoming a pretty tough chore. Toward the last of April, the snow became soft, so we did our traveling early in the morning and late in the afternoon. After leaving Whitefish Lake cabin, a funny incident happened. The snow had gotten really soft and travel had become more difficult by the day. The cabin was built along a big sand range with several sinkholes in the area, some as many as sixty feet deep.

From near the cabin, we picked up a ten-foot canoe which Red strapped to the top of his already heavily loaded toboggan. Each of us had on everything a trapper would need, and it was heavy. Since we weren't short of dog power, this fact wasn't bothering us.

After leaving the cabin, I took the lead, but hadn't made it more than 300 or 400 yards before I came to a spot where the only snow left for a trail went right alongside this big sinkhole. With some extra caution, I managed to make it past.

Coming along behind me, Red wasn't so lucky. He hit the edge of the sinkhole and the canoe tipped. Down into that blasted hole rolled radios, pots, pans, and what have you. All the way to the bottom of the sinkhole. Red was so mad for a little while that I thought he'd explode. In a way, he tried to blame me, but I'd have none of it, telling him he was driving his team same as I was and should've been more careful. It took us a while, but we retrieved everything and got underway again.

In a lot of places, the snow had melted from the tundra, so we had to pick our route carefully to get the sleighs through. Back at Sandy Lake, I saw a sure sign that we'd have to hurry to beat the thaw. The water was open at the end of the lake near a creek and two ducks flushed as we went by. We were going to be caught in between if we didn't hurry.

One day before reaching our destination of Beaverhill Lake, we camped at a trapper's camp from the previous winter. There were some caribou heads there, making it unnecessary for us to kill a caribou to feed the dogs. It was here that we began to worry about the Thelon River, and whether it would be frozen enough for crossing. It was really getting warm and our troubles were compounded by a chinook wind. We fed the dogs up well that night so we could make haste in heading for the river. When we discovered that the river was still tight and able to be safely crossed, we breathed a sigh of relief.

Red and I went to Beaverhill Lake on the Thelon with the direct purpose of building a cabin. Along with Phil, we planned to trap this area the following winter. A cabin was obviously more comfortable as a main camp than a tent. The whole idea had been Phil's. He'd already told me where and how it was to be built. Like most men in the early days, I'd had my hand in building a number of cabins, so building one wasn't a new experience. But doing it in such a remote place was.

When I'd asked Phil where he wanted it put up, he'd said right in the bloody open, well away from everything. He said he wanted the door on the west side, but I questioned his judgment. I thought the door should go to the south, but he wouldn't budge.

Phil and Red had pulled the logs to the site during the last winter, so things were ready for the work to begin when we arrived. We pitched our seven-by-nine tent and, with our radio and enough supplies to last until July when Phil was to fly out with a Norseman to get what we'd need for the winter, we had all the comforts of home.

So it was that we went to work like a couple of beavers. We had a little trouble getting meat for the dogs until after the first of June. A few caribou were around, but they were unreliable. You'd see them one day and they'd be gone the next. We dealt with the situation until July 1, at which point we moved the tent to Beaverhill Lake which had open water in the bay out to the islands. We put the nets out and started fishing. From there, we alternated, one of us fishing and the other working at the cabin.

We moved our camp at night because the flies were really getting bad. We went along a sand hill on the muskeg and a kind of flat. We traveled with the dogs, but had an awful time hooking them up. Once we did, they really gave us a ride to avoid the flies. Red was ahead and made a big mistake by stopping. I never saw so many mosquitoes clouded around a man in my life. Every time we wound back to this muskeg, a big bunch of them came up. We had to keep going, and so made that distance in no time. Surprisingly, there was a breeze when we reached the lake and the bugs had disappeared.

Near our camp at the end of the lake was a big freight canoe left by some trappers before us. Red would spend a day and a half with the dogs, tending the nets while I worked at the cabin. Then we'd switch jobs. The cabin took shape and was slowly finished. It was big, seventeen by nineteen, with a large veranda over the door.

The dogs were staked by the lake close enough to reach the water, but far enough back to escape the waves that were the result of the strong wind. One morning, they woke me with loud barking. Following their eyes to two small islands out in the bay 300 or 400 yards away, I could see caribou. They'd swum across from the mainland to the east and the islands were covered with them. The wind was quite strong that day and they were probably tired.

At about sundown, one cow came to the end of the island and tested the water. Pretty soon—and, boy, it was a beautiful sight—she went in, heading toward the mainland a mile away. The rest followed and, by the time all were underway, there was a black streak all the way to the main shore. The water was rough and part of the time the line was broken because some would disappear. They swam like a bunch of fish. I was without a camera, but I don't need a picture to recall that sight. These moments kept a man going out there. This kind of thing that was found only in the splendor of nature always moved me. No mechanical movie can do the same thing as viewing it right there, firsthand: the real thing can't be replaced with a thousand pictures.

For about eight weeks from the first of July on, we suffered a scarcity of supplies. Phil had planned to bring supplies in July, but he didn't show up. We waited and listened for the sound of the plane, but didn't see any other human being all of that summer. Once, we heard a plane well to the south over Beaverhill Lake, but we later learned that the government was mapping the North by taking a photographer in a plane over the area to take photos. Even though that wasn't much civilization to look at for a whole summer, that's the way I wanted it.

Phil didn't show up until September 13—two months late—but we made out pretty well with what we had. I became amused at Red when he ran out of cigarette papers. I thought for a while he might go crazy. Like a mad dog without water, he ran around trying to roll cigarettes from everything he could find. Nothing worked until he thought about using the paper around tea bags. It didn't do a bad job and seemed to satisfy and settle him down.

I had plenty of tobacco, you can danged sure bet on that, and since a pipe was my smoking weakness, I had no worry about papers. I wasn't a bit concerned, but it sure made me laugh to see him fumbling around trying to roll his smokes.

For those two months we lived on fish, bannock, and tea. Everything else was used up and those three things constituted the only choices available. We had little trouble catching plenty for ourselves and the dogs because Beaverhill Lake is loaded with trout and whitefish—and some big ones, too. Most lakes out in that country are like this, with most seeing a net or hook only very infrequently. If a man starves to death out there, it's because he's either too damned dumb or lazy to go fishing.

By the Jesus, though, after several days of eating the things three times a day, did I get tired of looking at fish—and I loved them, too. Still, by suppertime each day, the appetite had returned and a fish started looking pretty good again.

Fortunately, we had plenty of salt and a few tins of molasses. A man sometimes gets a craving for something sweet, and the molasses was all we had to kill that desire. Whenever the urge came, we'd go to the tin and dip our spoon in for a taste. It was that real heavy kind—thick and black. It made me shiver some while I ate it, but the stuff went down and took care of my sweet taste. Anywhere else, I couldn't eat it that way, but that's all we had.

I enjoyed the work and staying out there in the wide open with all the sunshine and fresh air. The thousands of birds kept us company as the area attracted them. Ducks and geese were also abundant. We always had companions because this area was a natural paradise, unmolested and safe from interference.

That summer was a wholesome experience and, physically, I've never felt better. The weather was very nearly perfection until August fourth, when we got a big surprise. All at once, it turned cold, the skies went dark, and a big snow and wind storm pounded us. Since this was unusually early and couldn't last, things cleared up as quickly as they had come. The snow melted and the warm weather returned to bless our stay.

We stayed holed up in the cabin until the storm blew itself to the south, coming out only to feed the dogs. When it did, I went down to the lake for a supply of fish and saw something that looked like vegetation on the edge of the water. I hurried to the lake for a closer look, knowing it couldn't be plants because Beaverhill Lake was danged clean and free of vegetation.

Once I got to the lake, one of the oddest sights I've ever observed showed itself. The dark mass I'd seen from a distance was bugs. That storm had killed every bug in the area, and the wind-blown water had washed them up along the shore. The result was a line of them about a foot across and maybe six inches deep, with some floating in the water and others plastered on the rocks by the shore. Since bugs were always bad in the Barrens until the frosts came, this was a blessing. There wasn't a single fly to annoy us for the rest of the summer.

The storm did its business out on the tundra, too. The caribou's instincts were stirred, and their migrating urges were set into motion. They started heading for the breeding grounds early. A very few days after the storm, we saw a cow running south like the mill-tails of hell. She was the forerunner. Flies always gave the caribou fits during the summer months and, with them gone, the herds were on the move.

The following day, the rest of them followed with all the purpose and dedication that are characteristic of this species during their yearly trek. It was impossible to estimate their numbers, with dozens of caribou on the move from every direction. This herd was at least seventy-five miles wide as it stretched out across the Barrens.

Red and I didn't sit tight for long. Having gotten over the shock of the early arrival of the caribou, we quickly set about in our pursuit of our winter's food supply. I had never gotten my caribou any quicker or more easily than I did this year. Better yet, since the flies were gone, there was no danger of the meat spoiling.

By the time Phil finally arrived in September with the supplies, all the work was completed: the cabin was built, the meat was cached, and the wood was gathered. We were sitting pretty with time to kill waiting for the trapping season to begin. Phil was surprised to find us loafing around with nothing to do. As a special treat, he had brought in some fresh steaks and fresh greens. We were loaded down with supplies, even a sleigh. And Red was happy as a kid with his fresh supply of cigarette papers.

After only a short stay, Phil flew back with the pilot. He was married and intended to bring his wife out to the cabin we'd built. We'd taken his dogs out with us in the spring with this idea in mind. But he had changed his mind in the meantime, and decided to try civilization and something other than trapping for a living. He told me later that the worst thing he ever did in his life was quitting the Barrens. I believe he meant it, too.

Red and I went through that entire summer with not so much as a single quarrel. We got along in most cases better than brothers. And we stuck it through the winter the same way. Each of us had his own line in a different territory. We weren't trapping partners, and we didn't split our catch. Each had his job to do and shared the cabin as a main camp.

Red was a fine trapper—as good as anyone who ever hit the Barrens. And he was good with dogs. But he wouldn't leave the cabin until I did. If I had my work caught up and stayed around the cabin a day or two to rest up, he also remained. It made no difference what he had to do; when I stuck around, so did he. Red was a friendly fellow and just liked the company.

With the exception of Joe Yanic who trapped west of the Thelon, we saw no one all winter. He was over a couple of times and I went to his camp one night. Red and I made a good catch, but white foxes were down to nine dollars and the quality of fur that year was very poor. More than that, the caribou were around all winter. They would smell the fish at a baited fox set and paw until the trap was sprung.

Still, it was a nice year on the Barrens. I really enjoyed not worrying about taxes or anything beyond food and supplies. The summer meant so much to me. After living it the way I did, it's little wonder that I can't understand how the natives can be happy in the settlements. That big country out there is meant for them, and I know they'd be happy in it. They'd have their freedom and they'd be where they've lived and died since history began. Sure, it's cruel at times. But the good things push the bad things so far back in a man's mind, he can't remember them.

Just the same, everything hasn't been perfect out here. For the most part, this area has been fairly clean of hard crime, even though trappers have competed for prime trapping spots and the natives have been somewhere in between. Since 1920, I can recall only two instances of murder out on the trap lines. But they were unpleasant events.

In the early days, two men from the States—Walter Remer and Bill Lind—were trapping beavers on a lake near the Dark River out from Fort Fitzgerald about fifty miles. Remer had quit trapping down on the Slave River and somehow hooked up with Lind after coming north. He had a wife and children, but they had gone south when he headed for Fort Fitzgerald. Remer was also quite a fiddle player, and he played for the natives at and around the post.

In the fall, Remer and Lind went out together as partners. But the following spring, Lind came back alone, sporting quite a large bump on his forehead. He went to the RCMP office and reported Remer missing. I knew the Mountie, Slim Bain, well, and had always respected him as a good man who had his job well in hand all the time.

Right away, Slim wanted to know the reason for his injured head. Lind blamed it on the rain, saying he'd slipped on the slick rocks on one of the steep ledges on his way over the portages. The police made a patrol out to the camp, and found Remer's dogs but not Remer or his canoe. Finding no evidence of foul play, the police let the matter rest. Lind stayed around the Fort Fitzgerald area for a couple of years, camping not far from me at the Buffalo Park. Phil—a fresh and ambitious seventeen-year-old—had just come into the area, and he got to know Lind pretty well.

One night, Phil came to my house to tell me some news I hadn't known before. He said Lind had told him that some people around Fort Fitzgerald were accusing him of having killed his ex-partner at their camp and hiding the body. Phil asked me if I thought Lind could have killed Remer. I told him the people blaming Lind were crazy.

Lind continued to stay around before returning to the States. On his death bed, he confessed to the Catholic Priest that he'd shot Remer in the back. They'd had an argument that turned into a fight. Remer had walked away from it, only to get plugged in the back with a rifle slug. He was then opened up with a knife, put in his canoe which was filled with rocks, and sunk in the lake. Parts of the canoe would come loose and surface years later, but the body was sunk forever.

After the Father heard the confession and Lind passed on, he notified the police. Once the man had died, he could tell the secret to clear up the matter. When the police received this word, they went for a look, but naturally could find nothing. Phil knew that lake, and he told me that it was probably 100-feet deep in places with some big cliffs around it. The police decided the two of them got into it over beaver skins, and the case was closed.

The whole thing was a surprise to me, because I had come to know Lind pretty well. He never once acted like there was anything bothering him. He carried that around inside him for those two years. That sure must've been a big burden for his mind—one I know I couldn't have lived with.

The other instance of murder happened out beyond Fort Reliance at the edge of the Barrens on the Thelon River. Jack Knox was in the country in the early days and, in conjunction with his trapping, worked with the Department, making patrols and keeping an eye on the musk ox around Artillery Lake.

One spring, a man from Ottawa by the name of Hoar came out. He was a minister, but also worked for the Department. At the time, I was on the Slave River and I met him as he went through. Boy, did he have a dog team: all identical big white Huskies. He told me he was headed for Fort Smith then on to Artillery Lake to pick up Knox, with whom he was going on east toward Baker Lake before coming back out by way of Churchill.

He got to Artillery Lake, found Knox, and, together, they drove their dogs to the Hanbury River. When they hit open water, they loaded their canoe and headed downstream. After getting to the mouth of the Hanbury, they built a cabin, and then made a series of patrols on the

Thelon Game Sanctuary to check the musk ox population. Afterward, they moved on down the Thelon River and eventually arrived at Baker Lake. There, Hoar continued on alone toward Churchill while Knox hired two Eskimos to go back with him to Artillery Lake. These Eskimos were brothers, one of whom was named Tekaluk. Somehow, the one brother drowned during the trip, but Knox and Tekaluk arrived safely at Artillery Lake and wintered there together.

The next spring, Tekaluk, who could speak very little English, told Knox it was time to leave. He said he was going any day and knew of an Eskimo man with two women. He was getting hungry for a woman. Tekaluk said he was going cross-country to eventually end up back at Baker Lake. Sure enough, Knox woke up one morning to find the Eskimo gone. Gone, too, were a 30-30 rifle, a bedroll, and a tea pail. Knox figured he was headed for the Eskimo camp located near the cabin where Jack Hornby had died. On the Thelon River.

We were coming to Fort Reliance from the Thelon in February for some supplies left there in the summer. Upon reaching there, we heard about the murder of two trappers on the Thelon River right at the edge of the game sanctuary and a little south. When I heard their names—Gene Olson and Emil Bodie—I couldn't believe my ears. I knew these men well.

When they went out, the police were told not worry. If they made a good catch, they would go out the other way to Baker Lake. If no catch was made, they were going to stay two years. As such, the police didn't worry about it when they weren't seen after the first year.

In the meantime, Howard Price and Jack MacKay, two other trappers on the Thelon, were scouting around for new territory. They were aware that Olson and Bodie had gone in during the summer a year or two before, but they never knew exactly where. While looking around, Price and MacKay saw in the distance a place that looked suitable for a cabin. Once there, they were surprised to find a tent whose appearance suggested that its occupants meant to return. A closer inspection told them something was wrong. They found the skeletons of seven dogs that had died on their chains. And the door of the ten-by-twelve was wide open.

Looking inside, they saw the forms of two bodies on their bunks. Some books were found at the back of the tent, one of which was a log book. Taking it outside for a closer look, they found an entry on the sixth of November of the previous year. It said they'd found Tekaluk the day before.

After Price and MacKay had come back to report finding the bodies, an RCMP inspector and his interpreter had made a trip out to investigate the crime. They came out from Fort Smith and went through Fort Reliance. Things had been left as they were found, so the inspector suffered no interference in his investigation. He determined that the two trappers had been killed with an ax. Besides that, one had been stabbed five times in the chest with a hunting knife.

When piecing together all the evidence and information, the inspector concluded that the Eskimo, Tekaluk, was the murderer. He thought it was done for a rifle, a .270 Winchester, which was missing.

As the log entry indicated, Tekaluk and the trappers had met in the fall. It would've been nothing for the Eskimo to kill them and then go on down the Thelon. Even though he had no canoe, he could've made a raft or a canoe. The natives could make canoes with caribou hides, no worry about that. Besides, he and the trappers knew each other from the winter spent with Jack Knox. He could've gained entrance because there was no reason for distrust. He might've done it, but I always wondered why he never showed up at Baker Lake. In fact, I don't think anyone ever heard of Tekaluk again.

Another thing, a trapper named Otto who came in with George Magrum, heard about the incident, and really became shook up about it, telling Magrum he was going to kill himself. The dead men were close friends and he couldn't take it. Magrum didn't take it very seriously and told him not to talk of such things. But, by God, Otto shot himself that night. The patrol was still out investigating when this happened and, when they returned to the scene, found reason to question Tekaluk's guilt.

They didn't know whether to blame Tekaluk or Otto. Later, it was proven that Otto had never gone within fifteen miles of Artillery Lake at any time, and it was some 150 miles to where these fellows were murdered. So, as far as I know, Tekaluk got the final blame for the crime. My only question is: where did the Eskimo go?

During the past five or six years, there have been some instances in British Columbia and the Yukon of trappers being murdered over trap lines and trapping territories. I've heard about these over the radio. But I don't believe this is what happened out there.

CHAPTER VIII

For the first six or seven years on the Barrens, my two brothers were fellow trappers. We frequently shared the same main camp and ran lines in different directions. A few times, we were apart by as much as a hundred miles or more, and only saw each other in the fall before going out to our territories, and in the spring, when coming in with our fur catches.

Hughie, the oldest, was a most determined cuss if there ever was one. He was a man who knew the trapping game and had enough drive for a dozen men. He was self-made and as honest as the day is long. And was he ever dynamite with dogs. If a dog didn't make double-time or refused to do his work, he was destroyed. We never completely agreed on this matter, and had a few minor quarrels, especially when my dogs were involved. But brotherly love prevailed and we always got over this right away. In all the years we trapped together, we never experienced any serious disagreements about anything.

The youngest of us was Phil. He was also a real trapper and loved the Barrens at first sight. He liked the excitement and adventure and was a tough bugger, too. And he was also good with dogs, hardly more than whispering his commands—and those dogs would hear him. I always marveled at this ability, because I had to shout at my dogs or they wouldn't pay any attention.

Both my brothers were good with their sense of direction out on the Barrens. Since I was always getting tangled up out there, the same couldn't be said for me. Those two guys depended on their lead dogs to get them around. They'd strike out somewhere and never seemed to get lost any time. A few times, they were gone for days on end, but they always turned up.

Every time the three of us came together, whether as partners or upon meeting anywhere, all we discussed was hunting and trapping. With us, the time never seemed to become stale as the thrill of the outdoors was buried in our hearts. We went out into the country to trap and make a living. We wanted to work at something we liked in a place where we felt free. And we did.

I believe that one who has the hunting and trapping fever and hasn't been able to go all the way has missed a lot in life. The three of us had this opportunity and lived with the game. Quite likely, we three partner-brothers should never have been made to change to other means of making a living, and stayed with the trap lines.

The love call changed all of this. It happened suddenly, with Hughie biting the dust first. Boy meets girl and that's it. He got married and started a trading post at Fort Reliance using the same cabin I live in today.

Phil stayed with me for a while, but the urge finally hit him, too. He went out and got himself a bride. But he hasn't forgotten the good old days and still hunts down his moose. Even though he has settled in Hay River, he traps yet, more for the kick than the profit.

For over thirty winters, I stuck with the trap lines on the tundra. Never, except for a few brief moments (and then not seriously), did I regret my job, because it was the one thing in life I wanted most to do. Yet, I wasn't alone for that entire time. The first few years after my two brothers hung their traps up, I kept at it by myself and was happy. And then I got this doggoned urge, too—to get myself a partner. It came about rather unexpectedly, but I didn't argue or try to avoid it.

In Snowdrift, an Indian settlement about sixty-five miles down Great Slave Lake from Fort Reliance, a widowed native girl lived. During trips back and forth, I'd gotten to know her pretty well while she was living with her first husband. His name was Lockhart and he was a good trapper. Like so many natives, he had died during a flu epidemic at a very young age.

This girl's name was Delphine, and a better worker never took a breath anywhere. Her life had been a tough one, since her parents had died around Great Bear Lake during an epidemic in the early twenties. A Chipewyan and no man's fool, she was quite young at the time, and was brought up at a mission from which she got a fine education mastering five different languages: French, English, Chipewyan, Dog Rib, and Slavey. She was an outstanding writer and read extensively.

While getting acquainted, we talked about many things, and discovered we had much in common. Finally, I came directly to the point and asked her to go to the Barrens and the trap line with me. She was alone at the time, and said sure, she'd go. Just like that, I had my partner.

I didn't have to quit trapping, either. And, I just as well tell the whole story and pull no punches about it.

We went out to the Barrens together, and lived and trapped as a partnership for ten or twelve wonderful years. Except for one year when we took a native boy out to cut wood, we were alone all those years. Delphine was more than a helper, and she gave new and truer meaning to the freedom of trapping. She was considerable help and a real source of companionship for me.

Several times I reminded her that she was silly to spend all her time in the bush. But she'd heed none of my suggestions to alter her lifestyle. This was her lifestyle, and she wouldn't have traded it for all the money in the world.

What a mind! She read every piece of writing she could get her hands on. All the classics, too. After we quit trapping and opened a trading post, people came out to Fort Reliance mostly to fish. It made no difference if it was a lawyer or anyone else, she could talk his language. Several times, she asked questions the experts couldn't answer. More than one man left scratching his head after an encounter with her, because of the way she could discuss so many topics.

The first couple of years, she was more help than I ever dreamed possible. At first I didn't take her out on the trap lines with me. Oh, she went along on the first trip when the traps were set, but after that, she stayed to wait for me at the main camp. I simply thought it would be too much for her to be exposed to the cold and blowing snow during blizzards and heavy ground drifts. It's sometimes tough enough for a man who is accustomed to it. Besides, I didn't want to take her along because I wanted to be by myself. I received a great deal of satisfaction from working alone.

This set-up lasted the first two years we were together, and seemed to be satisfactory to both of us.

Then came the third year, and everything changed. As had been the case from the beginning, we were at Beaverhill Lake near the Thelon River. In fact, all our years together were spent there at the cabin Red Noyse and I had built. My trap line was set in two one-day lines with a half-day line beyond. That means my traps were strung out at a distance that would take all day to reach my first out-camp. This would be either a seven-by-seven or a seven-by-nine-foot caribou skin tent with a canvas tent stretched over it. When I would reach this tent on the first night, I'd sleep there.

The following day, I'd go beyond a similar distance running traps until reaching a second out-camp and spending the second night there. These camps stretched a long ways to take advantage of the scattered patches of ground spruce. Without wood for fuel, this would've been an impossible task.

On the third day, I'd get up and run the half-day line beyond. That is, it would take half a day to run the line and the other half to return to my second out-camp. In the fall, I left some extra traps at the end of this line and set beyond as the daylight extended later in the season. By the end of the fifth day, I would be back to the main cabin.

Some years, the line would extend only one-and-a-half days one way with a one-day line in the opposite direction from the cabin. Covering a large amount of territory was important to success, because it was impossible to predict exactly where the white foxes would make their big run. I saw all my traps every ten days to two weeks.

The third year with Delphine, I was using a green lead dog, trying to break him to the degree of dependability required for my own safety. One Monday morning, I struck out over my line as usual. This dog wasn't familiar with the route and, to make a long story short, he got me lost and the weather snowed me in. That bugger took me a good three miles off course toward the Thelon River, and I had no choice but to lay out the storm in a snowdrift. Another thing, some of the traps I had at the end of that line haven't been located to this day. I never could find them because that dog was to be my guide. I had no idea where to look and that guldanged dog didn't know the way. Any good lead dog who'd been over the line before would've taken me to every trap, then home. Not this guy.

Well, we stayed lost for a time, too. When this big storm hit, I could do nothing but sit tight. Fortunately, I had my eiderdown and enough food to stay out of immediate danger.

The cabin had a big window facing east, and Delphine stood by that thing all the time it was light enough for her to see. On my return, this is the direction I should've taken. Instead, I came in from the west. My God, was that woman scared. That was on Saturday, and I'd been fouled up for nearly a week.

That little experience was too much for Delphine and she put her foot down. She got so she wouldn't stay by herself and went with me over the lines from that time on. Since there were normally seven good dogs and a good sleigh with plenty of power, she was never a problem. When it was time to leave for a run, I'd tell her I was going to strike out. By golly, she soon had things and herself ready. I couldn't get away from her. She was good, though—never a burden because she always did her share. After arriving in the evening at a camp, which might either be the main cabin or a skin out-tent, she'd have the fire going and supper ready by the time the dogs were out of the harnesses and chained for the night.

In the morning while harnessing the dogs, I'd look on ahead and there she'd be, maybe a mile up the trail. She was usually a bundle of activity. At times, though, while at the main camp, it was difficult to get her in the mood for traveling. She loved the cabin and, like the woman she was, hated to leave what she considered her home. She'd stall, wanting to remain for another day or two. Weather permitting, I'd tell her it was time to go, regardless of personal preference. When the weather was right for traveling, we had to go—that's all there was to it. The notion that Old Man Weather waits for neither man nor beast has special meaning for those of us in the Barrens.

When the snow was heavy in the spring, I'd frequently go on down the trail ahead of the dogs to break trail. Inevitably, I'd get only forty or fifty yards ahead when I'd hear her shout that I was bearing too far to the right. I'd growl some, but listened, because she was always correct.

With this natural tendency to bear to the right, I had to pay attention to her. Her instincts were sharply honed, and they caught me quickly enough to put us back on the trail. Many times, I'd have gotten us tangled up if she hadn't known. Her ability to make correct routing decisions was uncanny. She could take anyone from Fort Reliance to the Thelon River without missing a single turn. That's 140 miles by plane, but a darned sight farther by dog team and sleigh when you have to zig-zag around to hit all the wood piles for camping every night. Her background in the outdoors gave her the edge needed to be absolutely efficient in a world where few white women could endure. And believe it, no better worker has ever graced the tundra than Delphine.

That woman was constantly busy. She had to be doing something all the time. Whenever it was light enough, she much preferred working out in the open air. She could cook anything a man wanted and baked a bannock that was out of this world, but she hated to wash dishes. Many times, she would disappear after our meal was eaten. This never affected me, since I didn't consider washing dishes a bothersome chore; I'd take care of it.

My life with Delphine was good. I was able to be out there alone where I belonged and wanted to be hunting and trapping, but still enjoy the female companionship that all men need. She was efficient in so many ways, and made life so much better for me. Her folks had made their living this way, and she knew the duties and responsibilities involved.

With all the care and workmanship of a professional, she made all our garments from caribou skins for warmth. She dried our fish and meat to preserve it with the patience and skill that was characteristic of her people. Not only was she able to perform these tasks, but others, too, like no ordinary woman could. Like webbing snowshoes. She was unable to make the frames, which we purchased from manufacturers, but in the evening while we talked, she could weave a complete snowshoe webbing. Obviously, all these little things were a great help, very practical and necessary for our comfort and survival.

After all the chores were finished, she'd spend the idle hours of the evening making mukluks—fancy work or silk work. A few of her mukluks were trimmed with moose hair which she had previously dyed different colors. She used only the medium-length hairs from the mane on the back of a moose, and so would have to sort them, and work them into little bunches that she then tied into knots. She would use these whenever she wanted to make flowers or other designs. With only a pair of scissors, she could work all of this into a beautiful pattern. This was amazing work, and I was proud of it. I was especially pleased with a pair of mukluks she made for me. I've worn them for years. Whenever the bottoms wore out of them, she simply sewed new ones on.

This type of work was her mainstay, and it made her famous as one of the best artists among the native women of the area. One of her specialties was making mukluks from caribou legs. Taking mostly the hides from the back legs, she dried them in the fall and used them for the uprights, sewing a moose hide sole on them. This footwear was warm, comfortable and lasted for years.

There were times when she became so intense that she worked until two or three in the morning. This was out at our camps in the Barrens by light of a lantern. In the summer, a certain number of tourists came out to our home at Fort Reliance and she couldn't keep up with the things they wanted made. She not only made the usual caribou skin garments and moose-hide moccasins, but coats and collars from furs. She'd put her price on a piece of work and all hell couldn't shake her loose. Her work took a lot of time, but once a project was started, there was no stopping Delphine. She'd stick with it until the job was completed come hell or high water. She could really decorate a pair of moccasins or a buckskin coat. That's a handcrafting art all in itself and she was one of the best with her hands.

Much of her work went out of Fort Reliance as souvenirs just for show. One man summed it up very well. He said he'd been coming into the North for years from the States and had looked all over for some truly fine native work and now had finally found it. Delphine took great pride in her work, and it helped out our income.

When it came to tanning a moose hide, she was one of the best. Getting the finished product was hard work, as I soon found out. I helped her all I could in cutting off the hair and fleshing the skin. Delphine used caribou brains in the tanning process by boiling them into a kind of mash. The hides were soaked in water with this mash working slowly into them. This was allowed to set for a day or two, and then was followed by a rinsing-out and thorough scraping.

Next, I'd help her pull it. We'd go round in circles shaping the hide. Then it was laid on the floor and spread out to dry. We'd tackle the thing again every hour, pulling some more. Right before our eyes, we could see the thing tanning itself. By nightfall, after six or seven pulls, there'd be a nicely tanned moose hide. The only thing left to do was smoke it. Delphine did a very clean and wonderful job of that, too. The hide took on a rich, brown texture and was used for gloves, moccasins, and mukluks.

She could make jerky meat from caribou as well as anyone, too. Not only did she do an excellent job of drying and smoking fish, but could darned sure set a net where they were. Since she had lived off the land before, knowing what to do and how to do it came naturally for her.

By the Jesus, she could shoot a gun with the best of men, too. She was a crack shot. If I happened to be off somewhere and caribou came around, there was nothing to worry about. If meat was needed, she went to work on them without any hesitation. When she lined up a gun, there was meat on the table. In fact, she would sometimes shoot the caribou on hunts while I came along behind to do the cleaning and cache the meat. We became a good team, better probably than two men could ever hope for.

She could relate to dogs in a way that always pleased me. Never abusing them, she would talk to them and treat them like they deserved to be treated. She was able to handle dogs so well because there seemed to be a deep understanding between them. She took the time it required to teach pups what they needed to know. She had a lasting patience and never kicked or beat them around. At the same time, she was firm enough for them to know she meant business.

As young pups, they were taught by her to take meat from her hands gently. Never rough or cross with them, she would just bump their noses lightly with the back of her hand until they learned to take the food easily from her fingers. After she finished training them, they would carefully muzzle a piece of meat from your hand. We never had any of this snapping and snarling business. They became gentlemen.

Countless times, I've become busy with some job and all at once missed her. Not a person to lie around and be idle, she would take off somewhere when there was nothing to do. It was nothing for her to take the team and find a patch of wood. Pretty soon, she'd come back with a sleigh loaded with firewood. Maybe it would be thirty below or even in the summer. It made no difference to her, because she was tough and impossible to hold down.

When berries were in season during the summer, she would take off to gather them. She would go a long ways for them, figuring the ones close to camp were made dirty by the dogs. She was particular about things like that. When she returned with berries, I got ready for a treat. She could bake a pie that would melt in a man's mouth. This is how her people operated.

In the mornings I'd sometimes make a stack of hotcakes. She would eat them occasionally—only if they were on the thin side. I liked mine just the opposite—the thicker and heavier, the better. Whenever she could get a trout head, she had her kind of breakfast. She'd split it in two, take it outside and barbecue it. That was her idea of a wholesome breakfast. That thing was only half cooked and the blood could be seen running through it. She wouldn't eat eggs or bacon if there was a trout head around. I never argued, because she was brought up that way. She insisted there was more nutrition in a trout head than in the rest of the fish.

There wasn't a lazy bone in Delphine's body as she was always looking around for something to do. She was great company and outstanding help in the years we spent together, and I was very happy during this time. We were both pleased with our life because we lived it as we wanted to. But all good things must come to a halt, or at least slow down, and so it was with my trapping days. The past ten or so years it became difficult for me to handle the dogs. The life became too rough, and so we started our little trading post at Fort Reliance.

Here, she continued to keep busy by selling small items—things like beads, thread, cloth, and other dry goods—to the native women. She drove a tough bargain when it came to trading, and did pretty well at it, too. The natives were hard to deal with at times, but didn't pull anything on Delphine. She told them how it was going to be, and if they didn't like it, they could go someplace else. She got along better with the women than I ever could.

She did a couple of other things in a big way after we quit the trap lines. She really went after the books, reading everything she could get her hands on. And she translated the gospel from English into Chipewyan. She had always been a good writer, and the priest at Snowdrift told me the work was to be printed into books and pamphlets for use by the Indians. Delphine made the priest's work easier by putting the gospel into words the natives understood. It took her two years. No, Delphine was far from being without intelligence.

All those years it seemed as if we had been hiding behind a screen. We weren't married, and it began to have an effect on her. She wanted to change all this, and so got after me to get married. Given that my first wife had died a couple of years earlier, I could find no reason not to, and so quickly agreed. We contacted the priest at Snowdrift and were married in our cabin at Fort Reliance. This was a real Northern wedding—quite a blow-out. The boys from the Department of Transportation weather station came over and we had quite an evening. Most important to Delphine, the ceremony took her from common-law wife to the real thing. After all those years together, getting married made a difference to her. Maybe it was my imagination, but it seemed like our friends were more cheerful with us. One thing should be clear—we knew the difference between right and wrong. There was no getting around that. But, as things went, we got along fine and were good for each other.

A couple of years ago, Delphine suffered a stroke that left one of her arms partially paralyzed and nearly useless. She was laid up for quite a while, but the thing appeared to be getting better. It seemed like she was going to overcome the effects and return pretty nearly back to normal. Then last winter she began to act rather strangely, almost as if she had a premonition that something bad was about to happen. She feverishly started reading the bible, and spent long hours poring over it. She had become quite religious over the years, but this action seemed to be very extreme. I would catch her staring blankly far away. She talked very little and was edgier than I'd ever seen her before. Through all this, I never questioned her mood. I figured she would tell me about it if she wanted me to know.

One time, I came up behind her quickly and threw my arms around her neck. Damn, she got mad. When I finally turned her loose, there was the demeanor of a wild animal in her eyes. She looked all around the room and reached for anything she could get her hands on. I knew she was trying to find something to throw at me—a knife or anything. That scared me badly, and I never did it again. She had never been like that before, so I let her alone.

This past April as we sat talking in the cabin, she told me she was going out on the trail leading into the bush behind the cabin to get a load of wood. There was no reason to object, so I told her to go ahead, but not to be gone too long. That was her way, as she wouldn't be tied down.

She dressed in her heavy caribou skin clothing, put her snowshoes on, and took off. The dogs were with her, as usual. With some chores to do, I became occupied and paid no attention to the time. One of the boys from the weather forecasting station came over and said he saw the dogs outside. He asked where Delphine had gone.

Immediately, I suspected something was terribly wrong. Those dogs were her friends, and they had come back without her. This young fellow went back over the trail and there she lay. She had another stroke, and there was no telling how long she had been on the ground.

Yellowknife was called from the weather station and a plane was sent for her right away. She was taken to the hospital and has been flat on her back ever since. Like a vegetable. She can't talk and doesn't know anything or anybody. They're feeding her through her nose and I'd stake my life on one thing: she will never see Fort Reliance again. This seems such an injustice and I hate it with all my heart, but what can a man do? Delphine isn't old, either. Only in her fifties.

I went in on the plane with her and stayed over a month. Every day I went to see her lying there in the hospital. No human should be like that. Especially Delphine, because she was such a free spirit and was never one to lie around. She was always active and it tore my heart out to see her like that.

After a month of standing helplessly around, I pinned the doctor down about what he thought her chances were. He told me there was very little hope and nothing I could do, so I flew back to Fort Reliance. She has been in the hospital in this same condition for nearly a year.

CHAPTER IX

Delphine and I never looked upon each other as being different in color or background. We weren't a white man and a native woman, but, quite simply, two people doing a job and living a life which was mutually loved. We were good both to, and for, each other. Each of our lives had a missing link and the other provided the companionship required to enjoy fully the years of later life.

Throughout the years we spent together out on the Barrens, there was never-ending excitement and adventure. Dull moments were quite few and far between. There were some frightening times, but we were seldom scared. We learned to respect things and were careful when difficulties arose.

I vividly recall an instance while near the Thelon River that caused a great deal of excitement and concern. The two of us were working on the wood supply and doing it as a team, as usual. Several hundred yards from the main cabin on a sand ridge grew a small patch of ground spruce. We were gathering our wood there, with Delphine on one side of the ridge, cutting brush and roots, and stacking it as she went. I was on the opposite side, loading some cut wood on the sleigh and preparing to haul it back to the cabin by dog power.

I wasn't really paying any attention to anything beyond my chore, but upon turning around and glancing down I became alert with a start. In the snow just a few feet away, was the print of a huge grizzly bear. My first worry was for Delphine, because one can never predict the mood of these big guys. I unloaded the toboggan in a flash, and took off, following his trail. It led in a direct line toward the spot where Delphine was working. I watched that bugger, big as you please, standing up on his back legs to watch her. Quite likely, the dog team had scared him enough to cause him to take off. There was no way to tell just how long he had stood watching her. When he left, the wide open country was his calling and we had no desire to take up the chase.

Later, when I told her about the incident she was more mad than frightened. She took pride in being alert, but this time had failed. From time to time, different people have tried to tell me there are no grizzlies around the Thelon River. I know better, and say the doubters are crazier than hell. They may be scarce but, believe me, they're out there.

There was another brush with a grizzly at Beaverhill Lake near the Thelon. One morning when little was happening, I decided to walk to the fish cache near the lake shore. We had done all our fishing near this spot during the last fall and had stored our catch upon some stages after first drying the fish. We checked this occasionally to make sure it was safe. The smell of fish attracted most fur bearers and fish was a favorite food.

Before we reached the fish cache, we passed another containing caribou. This was also a stage. It was late in October, so there was snow on the ground. To my surprise, the stage was thoroughly wrecked and the tracks of a big bear were everywhere. Some of the prints were old, but most had been made quite recently. At the time, I was carrying the .250, which was a fine rifle, but with that old grizzly around, something with more power would've felt much more comfortable.

With the stage torn down and in shambles, the caribou meat was almost non-existent. The bear had dragged it all away except for a head, which lay on the ground a few feet away. Making several trips to complete the job, the bear had worn a regular trail leading back to some ground spruce along the ridge. I felt a prickly sensation and knew that big devil was close at hand because I hadn't seen his tracks to the west as I came down the trail.

I began to wonder if it would somehow return for that head left behind. Thinking it was a strong possibility, I jumped atop what was left of the stage. It soon became obvious that it would be foolish to take any unnecessary chances because the bear would consider this place his private food source. Looking around to all sides and seeing nothing unusual, I let out with two or three loud yells, figuring if the bear was close, he'd come out to investigate. I might just as well have saved my breath, because there was no sound or movement. I considered this to be very unusual, so climbed down, keeping my eyes wide open, I assure you.

I knew enough about bears to be sure of one thing: if that brute came out of his hiding place, it would head my way since this was the fall and bears aren't afraid of anything at that time of year. On the other hand, during the spring season he'd probably go the other way.

After getting down, I made a large circle in an attempt to find his tracks leading away from there. I hoped, but in vain. One thing was for sure: I wasn't about to go into the area where the bear had dragged all that meat, because I could just visualize him asleep right beside it. I didn't want to surprise that creature in the least. Keeping on with that wide circle and arriving at the ridge beyond, I found no bear signs. I climbed the sand range and studied everything through the rifle scope, and still saw no bear. This was very strange, because the country I surveyed below was open. I kept going with the completion of my circle toward the main camp, and still didn't see a single bear track. This was proof enough that the bear was still somewhere close to the caribou cache.

While all this was going on, Delphine had taken a little trip of her own in the afternoon. By this time of year, the caribou should have been well to the south a good ten days, and neither of us had seen any recent sign to indicate otherwise. Delphine had gone out carrying her 30-30, intending to retrieve a pack sack forgotten the day before.

The pack sack was found alright and, on the way back she sat down on a gentle rise right out on the open tundra to roll a cigarette. While having the smoke, she looked to the left, and spotted six big bull caribou coming right for her. It was a clear, windless day, and even though they saw her, the caribou neither smelled nor recognized what she was, and so kept coming. When they got abreast as they usually did under these circumstances, she was able to plug five of them. The last one only got away because she couldn't reload in time. They fell only a few yards apart, as I found out when going out to bring them in. For stragglers, they were fat and in excellent shape—ideal because they were needed for our own eating meat. Ample caribou were already cached, but had been killed early for dog food.

When Delphine came back to the cabin that evening, her misplaced pack sack was filled with caribou fat for cooking. As efficient as ever, she had gutted and turned them belly down so they'd be in an easy position to load onto the sleigh and pull in with the team. They weren't cached on the spot only because there were no rocks and brush around for that purpose. She couldn't have gotten them in more open country.

She was proud of her successful hunt, but sobered up when I mentioned the big grizzly that had broken down the stage and carried the meat away. Deciding we couldn't share our hard-earned meat with a rampant bear, we planned to go after him. With our food supply threatened and no way to replace it, we could take no chances. We arose the next morning at the streak of dawn, and went down toward the lake in search of the destructive grizzly. Delphine had the .250 ready to go and I was armed with the .270.

Once more, we made a big circle around the cache without spying a single bear track leading out of there. After studying the mass of tracks around the stage, we decided the bear had been in the timber patch for at least two or three days. I was quick to note that the caribou head was still where it had been the afternoon before. Standing near the stage, we made plans to go in after him. I gave her final instructions to stay no more than thirty or forty feet to my right. If she spotted him first, I told her not to wait on me, but to start shooting, immediately.

In we went, ready for anything, and had put no more than a dozen paces behind us when she called out. Pointing ahead, she indicated the bear. But she said he was already dead, and didn't raise her gun. I told her she couldn't be that sure, but she came right back, telling me the bear was bloated, and stretched out with his legs at his sides. That's when I saw that son of a gun, not more than fifty yards away. I had to agree that the bear looked dead, but after a few more steps, I lost my nerve and shot it. The animal was bloated and filled with gas. I shot it through the shoulders and when the slug hit I heard this loud "psst." Only then was I reassured that the bear was dead. But I could find no reason for its death, as there wasn't a sign of a shot other than mine.

Unusual at this time of year, it was thin as a rake. A big she-bear it was—a Barren Land grizzly, and a huge one at that. Delphine helped skin it and we kept the hide and skull. At that time, I was on predator control for the government, and so turned the hide and skull over to them when they came out on patrol later in the winter.

After skinning the bear, we cut off a couple of quarters to use as dog food, even though there was plenty of caribou and fish. In the Barrens, meat was never wasted if it could be prevented. Anyway, we left the rest of the bear where it had fallen, and it lay there all winter. I frequently checked around it for fur sign but, in spite of a surplus of wolves and foxes all winter long, they wouldn't touch that bear carcass. Barrens animals will normally eat anything, so this was a strange incident indeed.

Working on predator control for the Northwest Territories government, I was given and permitted to use poison in baits for the purpose of keeping the wolf population in check. Figuring the bear to be good bait, we took her insides out and I anticipated getting some wolves from them. After hauling the paunch out, I wrapped it inside a caribou hide and fixed it up with poison. I checked that bait all winter long and it didn't yield a single wolf. They wouldn't touch it.

One day, we drove past the bait and saw a set of wolverine tracks coming from a burrow beneath it. That little devil had been in sniffing around, but had pulled out without so much as a bite. Looking out across Beaverhill Lake, we could see a little black spot. It was the wolverine, which had stopped out there and was glaring back at us. The dog team saw him, too, and was eager for the chase, but the sleigh was heavily loaded; we could never have caught him. Taking the shovel and digging down to and around the bait indicated that the wolverine would survive, because it hadn't touched a scrap. Wolves also prowled around it all season long, but, for some reason, stayed clear of it.

Spring came and we went out to the South with our catch. The following fall it was back to the Barrens for another trapping season. Immediately upon arriving back at Beaverhill Lake, I was anxious to see if the wolves had eaten the meat from the bear carcass. Because we'd only removed two quarters for dog feed, plenty had been left to tempt them. It wasn't fat, but, still, it was meat.

But when I examined the carcass closely, I was surprised to find that the only thing that had paid a visit to it had been some gulls, who'd picked at the meat between the ribs. The sun had dried it and everything was still there. This was very unusual out in the Barrens, because wolves following the caribou back to the tundra in the spring will normally eat anything.

The conclusion I made concerning the bear's death was that it had gorged itself on the caribou meat while in a state of starvation and died from the effects. I've seen this happen to dogs and there was no reason to believe a starving bear couldn't suffer the same fate. A dog, when it hasn't eaten for a long time, will eat so much he will kill himself. This happened long before the trapping season with no possibility of the bear finding any poison. And there were no bullet holes. That danged thing was poor and starving, and died from overeating.

This and other unusual occurrences were part of our life together beyond the trees. Delphine was made for the country, having lived her entire life in its surroundings. I had adopted it and felt that it did the same thing to me. The two of us were destined to spend our lives in the North, and circumstances had brought us together as one. Ours was a good union—one that was right for both of us.

In the spring, we came back to Fort Reliance from Beaverhill by dog team, a trip which, at times, was a mean one. It was perhaps this trip above all other events in our lives together that best shows Delphine's heart and strength.

It was nearing the first of April, and with all our clean-up chores completed, we were ready to close our winter operation and head in with our furs. The sleigh was loaded down with plenty of dog feed to last until Fort Reliance. We had caught a lot of fish and it was part of the dog feed: green, dried, and fresh. There was plenty of everything to keep us going for the whole trip.

A trapping friend of ours, Jack Knox, was working the same general area about thirty-five miles to the west. His camp was nearly on our route, so we decided to drop in on him with the thought of traveling back to Fort Reliance together. Even though I had been to his cabin several years previously, I wasn't positively sure of its exact location. We hadn't seen Knox all winter and were anxious to be off for our visit with him.

Just before we were to leave our cabin, a police patrol stopped in. The man guiding the Mountie hadn't been out to this country before, and the two of them had one awful trip. They were worn down and in pretty bad shape. The situation was made worse by the fact that they had found no caribou. Their dogs were in sorry condition, with starvation only a few days away. There was no alternative but to give them much of the meat which was intended to get us by on the trip to Fort Reliance. We took the fish off our sleigh and fed their dogs. The patrol stayed around for two days before getting straightened out enough to travel. And Delphine and I came up with a meat shortage because of this encounter.

The patrol had also intended to stop by to see how Knox was getting along. When we were at last able to get away from our cabin, we headed west on a line with two skin tents in which we intended to camp. But the patrol was never seen by us again since a howling blizzard meant we got stuck at one of those tents for two nights. This was one of those bad ones where we couldn't see a thing. Unfortunately, the delay cost us a great deal of our valuable dog food. The scarcity of meat was becoming an emergency with the caribou still to the south in the timber. Our chances of getting additional meat were very dim indeed; although we saw a few signs of caribou heading toward the tundra, nothing seemed to look recent.

Eventually, the storm let up, enabling us to continue toward Knox's cabin, which was really twenty miles off our course had we intended to go directly to Fort Reliance. As luck would have it, we couldn't locate that cabin. With the temperature bitterly cold—forty to fifty below zero all the time—we followed a ridge, staying on it until turning on a line headed for our old trapping grounds, ninety miles east of Fort Reliance. There was an old cabin in this area where we knew some flour and baking powder had been left for emergencies, and we figured to use them. If we couldn't find some caribou in the meantime, the very least we could do was bake some bannock for the dogs.

Along this ridge, we got stuck in another danged storm before leaving the open tundra. The only thing to do was pitch the tent and hole up again. During this storm, we even had to feed our own eating meat to the dogs. It was at this time that I really began to worry about our predicament. We were carrying a .410 shotgun along with us for the purpose of shooting ducks and ptarmigan. But in spite of the fact that there were ptarmigan tracks everywhere we looked, there were no ptarmigan.

One evening when the storm had eased up a bit, Delphine went hunting for these elusive birds and succeeded in killing two of them. We still had a small amount of rice, some caribou fat, and a little salt, thank God. Delphine boiled it all together and I believe this was the best feed I've ever had. I was hungry to the extent that I was taking vitamin pills and trying to live exclusively on them. I had taken too many, because my limbs and joints were becoming stiff, so I quit taking them and felt better almost immediately.

This storm finally passed on through, enabling us to move again. The same ridge we were following finally led to a lake I felt had to be Whitefish Lake. The cabin we were seeking was on this lake and, after a short search, we located it. We were again happy because the flour and baking powder were still there, along with some fish.

The police patrol, as we learned later, had found Knox's cabin, but failed to locate him. Instead, they found a note stating that he had left the cabin on the third of February headed for Fort Reliance because he'd hurt his foot in some kind of accident. This was the first year Knox had been in this particular area, although he'd trapped for years around Artillery Lake, over 150 miles north and west. When the police found no one at the cabin, they returned to Fort Reliance to report Knox missing.

Because of the missing trapper, search planes were sent out to tour the country. Just about a half mile before Delphine and I reached the cabin on Whitefish Lake, we saw a plane coming from the west. While it circled over the cabin, we immediately turned the dog team around on the lake so the pilot could see us. The dogs were all blacks or black-and-whites spread out broad-side so there was no reason for him not to see us. I even took off my parka and waved it. Still we weren't spotted. The plane went on toward the west and we never saw it again. Later, we learned they were searching only for Jack Knox.

That day, Delphine froze her face as we drove down Whitefish Lake toward the cabin. I was trotting ahead of the dogs on snowshoes when the first hint of her trouble came to my attention. The cold was breathtaking in its bitterness. Once or twice while stopping to let the dogs catch up, I noticed she was crying. When I asked what was bothering her, she said her face was cold and smarted like it was on fire. Thinking this was not unusual or overly serious, I told her to try talking more and maybe this would prevent her face from becoming frostbitten. Her clothes, including a hooded parka over the top for added protection, were very warm. It never dawned on me that this was to be a terrible experience.

It happened nevertheless. That night at the cabin we did all we could for her poor face. Just looking at it told the story. I felt sorry for her because it had to hurt like hell. Over the years, the scar stayed with her as a permanent reminder of that day. It gradually faded, but anyone could tell it had been frozen—her lips, chin, and the skin above the lips. It was a gruesome scar. We went to several doctors who were able only to prescribe some pills and salve. Nothing helped except time.

We stayed at that cabin on Whitefish Lake for one night, feeding the dogs heavily for additional strength. There was even half a case of pilot biscuits, which the dogs got, also. They had become weak from not eating. One of them, Rex, had staggered so badly in the morning that I had to help him stand before he could be placed in the harness. Once we were underway, he had worked out of it and seemed to be all right. Every member of the team was a fine worker—and had to be, because there was a huge load on the sleigh: all our gear and fur catch, along with the tent and stove. They were good dogs, and it was nothing for them to pull the load. In fact much of the time, both of us rode on the sleigh, too.

After leaving the Whitefish Lake cabin, we still had some eighty miles to cover before reaching Fort Reliance. We had to first cross Whitefish Lake, a distance of nine or ten miles, to get to the narrows leading into a string of lakes we called Sandy Lake. We knew the location of a cabin and headed for it.

A heavy ground drift was on when we left, but this was no big difficulty because we could see well enough to travel. Just before reaching the end of Whitefish Lake, Delphine, who was walking ahead of the dogs on snowshoes, quickly came back toward the team. Whispering very quietly, she said there were caribou ahead of us. Thinking she might be seeing some timber or imagining it, I asked if she was positive. She insisted they were caribou because there was movement and it seemed to be right toward us. It was so easy for a person looking through all that blowing snow, to have the imagination play tricks on the eyes and mind. It didn't take me long to remove the rifle from the sleigh—and to do so quietly, because the dogs would've caught on and become excited. She whispered "caribou," because the dogs knew full well the meaning of the word.

Walking ahead carefully with the .250 ready, I hadn't covered more than 200 yards and there they were: eight big bulls coming straight at me. When I got down on one knee to shoot, they were almost obscure in the ground drift. I fired four carefully aimed shots, but could only see one caribou drop. I felt good at the time, knowing one was better than nothing. Delphine had done her job well, as the dogs were held tightly by the head line.

I was very pleased while trotting over to the dead caribou. To my complete surprise, there was not one caribou, but four. Unbelievably I'd downed one with each shot. The gloom of the blowing snow had hid the results.

Taking the insides out of them and putting all the meat we could on the sleigh, we struck out for the wood patch at our old camping place to the north at the end of the lake. We hurried and it was fortunate we did because the weather became very dirty just as we arrived. A spring blizzard was once again about to engulf us. By the time the tent was up, it had become bad, with the wind screaming around us.

I was able to get out long enough the following day to retrieve the rest of the meat. We were held up by that storm for three days. During that time as the meat was dried, we stayed busy. We set poles up inside the tent with meat laid across them. I kept the fire below it burning. Even though it was bitterly cold outside, we didn't care; we were a completely happy couple. The dogs—suddenly in possession of plenty of meat after a long spell of hunger—didn't mind, either.

Not leaving a scrap of caribou meat behind, we found the fourth day fit for traveling. We made tracks for Sandy Lake, which wasn't very far away. Although I'd been this way several times before, the country didn't appear completely familiar. I knew the sand range we were traveling, but couldn't decide if we were traveling on a muskeg or a lake. Delphine went ahead for a ways, but didn't fare any better.

We were looking for that cabin and it was getting late. I thought it possible that we were in the wrong valley, so we cut back a little to the north. Pretty soon, we ran into some danged rough country and I knew the choice had been wrong. Returning to a patch of wood we'd passed earlier, we pitched the tent to wait for another day.

Since there was plenty of food, we weren't in any danger. Still, if a man had his choice, he liked to know where he was all the time. The night passed comfortably and, by early morning, we were again ready to move. Delphine went on ahead while I was loading the sleigh. Before she was out of earshot, I heard her yell. She had found the cabin.

Why we hadn't noticed it the night before, I'll never know. It had a good stove and all the comforts of home. By the Jesus, I felt like camping there, anyway. But we grudgingly passed it and continued to the end of the lake where we spent two more days and managed to get two more caribou. We now knew exactly where we were, and had plenty of food. The memory of those three days of starvation was forgotten, and we felt like kings.

Taking our time, we fared quite well until we were about twenty-five miles from the tree line. Here, we somehow went into the wrong draw and got tangled up again. Delphine was determined to work toward the south, but I insisted on going to the northwest to hit a big mound. All the time, we should've paid attention to the lead dog, because he knew the route. But I thought I knew more than he did, and wouldn't let him have his way.

To complicate matters, a thaw lasting for three days hit, and it prevented us from going anywhere, even on snowshoes. The entire bottom dropped out of everything, leaving us no choice but to camp. It became cold enough to once more tighten the snow, so at five o'clock in the morning, we were off again. The sun was up as the days become long in April. Since it was a very calm day, sound carried from a long distance. We heard a plane taking off from Fort Reliance, one searching for Jack Knox who still hadn't turned up.

Deciding to attract the pilot's attention, we cut a big pile of spruce and set it on fire. We accomplished this in short order because the plane seemed to be heading directly at us. It swerved a little to the south not over a mile away, but there was no indication that our fire had been seen. There was enough smoke to draw the devil out of hell, but it didn't attract the plane. It went on toward the east looking for Knox and—as we later learned—for us, also, because we were overdue.

With the plane out of sight, we resumed our journey and soon spotted a bunch of ravens circling over what had to be a wolf kill. Even though it was about a mile off our course, the dogs were chawed in that direction. Sure enough, a big bull had been pulled down in deep snow as he'd tried to make it to the lake. All that remained was a stack of bones with some heavy meat along the upper legs and around the head. We put the whole thing on the lazy back and took it with us. The trees weren't far away and, shortly, we came across some birch followed by bigger spruce. This indicated that Great Slave Lake was very close, even though we were off course.

We made camp that night and broke up the bones, saving the marrow. There was plenty of meat, but marrow was highly nourishing and offered a change. The next day marked the start of the last leg of our trip. After crossing Pike's Portage and coming into Great Slave Lake, there were still nine miles to travel through a lot of water on the ice.

When we arrived at the barracks, we were completely worn out. My first question was about Jack Knox. He still hadn't turned up. They were to the point where his equipment—his canoe, outboard motor, and various other bits he'd cached there in the fall—was to be sold. I talked them into holding off because I felt that, unless there had been a serious accident, he would eventually show up. By golly, on May 2, that old bugger came staggering into Fort Reliance. He explained that he'd gotten too far south and had shot all his dogs for food. All he had for traveling were his snowshoes. With the thaw coming, he couldn't go very far each day. He lived on what few berries he could find along with a few grouse he'd been able to shoot.

Surprisingly, he was in pretty good shape. When he walked into the barracks, the Mounties thought they were seeing a ghost. Knox had been missing for over eighty days. Instead of asking for food or comfort, the first thing he called for was his mail.

CHAPTER X

In my world beyond the tree line, a man could walk, use a canoe, ride a plane, or follow a dog team. The mere mention of one of those new machines—a snowmobile—will attract a lecture with a pointed finger from this old-timer. That risky new contraption doesn't belong out on the tundra, either as a conveyance for trapping or anything else.

Back in the early days, the canoe provided the only available means for getting into the country, and was of vital importance to all trappers. It was also capable of carrying all the stores needed for the winter. A trapper would either leave the south early enough to paddle and portage it close to his main camp, or cache it along the way when the snows came, and let the dog team complete the trip in.

The coming of the plane greatly diminished the importance of the canoe, but never completely eliminated its importance. Still, the greater advantage of flying to the trapping camp versus canoeing to it were obvious, and that's why many took the easy way and flew out to the Barrens from Fort Reliance. The big conversion to flying by the trappers came about in the Northwest Territories during the early thirties.

Down through the years, whether by canoe, dog team, or plane, the lure of caribou, wolves, and white foxes pulled me to the Barrens for my work. Once settled for the winter, my trapping success, return to the South, and very life were directly dependent upon the dogs. All those modern things are wonderful, but without good dogs, the Barren Land trapper could never have come into existence.

Never in Christ's world will the snowmobile or anything else replace the dog as a man's means of transportation on the tundra. For one thing, how could anyone find companionship with some damned machine? As a man living alone, I got a kick out of the dogs. They became my friends, and were the best danged company in the world. Watching every single move I made, they lived right with me and wanted to please all the time. I treated them right because they were my friends. Try that with a machine.

Unfortunate things did happen from time to time, and a dog was lost either by accident or sickness out on the trail. I would miss him and always feel sad because of it, but I would still make it to camp that night because there were five or six left in the team. If I'd had a danged snowmobile and one part of it went haywire, well, more than one person has had to walk back from a trip. Out there, it wasn't just a mile, either.

Red Noyse lived at Reliance for a while, and bought a machine from one of the boys at the weather station. One day, he took it out and went way the hell and gone across the lake to the opposite shore. That was probably twenty miles away, and the danged thing broke down. A rubber or some fool thing fell out of it with oil going all over the place. Red was stuck, and had to walk all the way back. Not on your life would that have happened with dogs. No, sir!

I have a suggestion for those hit by the snowmobile urge: Get everything loaded and set to go, then hook a cariole on behind and load in five or six dogs. They may be hard to handle with all the noise and speed, but when that damned machine breaks down, they'll return you safely home. They'll probably laugh and snicker all the way back, too.

And there's another important thing about this rivalry between dog and machine. Suppose a man elected to take a snowmobile with him to the trap line and his traps became snowed over? Or his trail became obscured by blowing and drifting snow? There's very little to go by out in the Barrens because it's just a big white expanse where everything looks alike. How the hell would this man and his machine find anything? With a good lead dog, there's very little to worry about because that bugger will take you there. He'll deliver you to the bloody sets no matter where they are. He'll stay on an invisible trail and come to a stop beside every covered trap because he knows his business. And having reached the end of the line, he'll take a man home, if permitted to do so.

There is no machine in the world that can do all this—plus keep a man company. Unlike a machine, those dogs were alive and warm and I did everything in my power to keep them happy. As long as I had my dogs, there was never any loneliness. We trusted and understood each other. I fed them well and kidded them along like humans, and they worked their hearts out for me.

In this country of dogs and dog men, I've seen all sorts of abuse directed toward these creatures. Men have called them dumb animals, and they're crazier than hell, too. I've gotten used to others doing it, but I never mistreated my dogs. If one of them came into the house, I'd yell at him to get out. But I'd never kick a dog in the butt to hustle him up.

Some guys curse and berate their dogs like mad men. I refuse to do this, too. If someone were to do that to one of mine and I was physically fit, I think I'd peel my coat off and go after him. Dogs are like human beings. All a man has to do is feed them once a day, then put them in the harness. Baby them along tenderly and they'll kill themselves for you. They will repay this kindness and love a thousand times. Show them some affection, take time to teach them, and they'll live their entire life just for you. Dumb animal? Not a chance.

I carried a whip but not for punishment and I snapped it above their heads to break up fights, or alongside the leader to make him leave the trail for a new route, but I've never struck my dogs. Others have not been so kind. I've seen dogs all broken up—ribs, legs or backs—many times. But if a man uses instruments on them like some of the natives do—many beat the hell out of their dogs using clubs and sometimes rifle barrels—problems won't be long in coming when this same man puts these same dogs into the harness and hooks them to a sleigh. There is no trust when dogs are half mad and biting at the native. And the dogs can't be blamed for their subsequent behavior; they're simply reacting to the way they've been treated.

In all the years since coming into the North, I've never been able to speak of dogs very long before thinking about some particular individual or special team. When recalling a dog of the past, my mood goes to a certain point of reverence most hold only for departed loved ones. But to me, that's what they were. It doesn't bother me to feel a choke of emotion when remembering a good dog or a sad event that caused the death of one of my team. My memory of the years gone by is full of experiences centered around these faithful companions.

However unpleasant it may appear, I've had to accept the fact that dogs were sometimes killed to feed other dogs, and I even once had to eat dogs or starve to death. This was solely through absolute life-sustaining necessity, done in the absence of any other choice and completely against my basic nature. And after it was all finished, and the crisis had passed, the degree of sadness I felt was impossible to describe.

A multitude of hours throughout the solitary life I chose were spent with only my dogs as companions. It helped to talk to and receive response from a listener, a dynamic that was fulfilled while talking to my dogs. A soft look, the turn of a head, the slight raising of an ear, or the wag of a tail was all I needed. The special reaction a well-treated dog reserved for his master was all that stood between me and loneliness. There is little wonder that, in my recollections of the past, dogs sometimes receive greater mention and stature than do some of the people I've known.

Since I treated my dogs as loyal friends, certain actions of others were difficult to understand. One time—and I know this to be a fact because those involved told us about it in Resolution—Gus De'Steffany and Jack Stark went well out into the Barrens to trap on the Coppermine River. They established a main camp somewhere around Lake Providence, either finding or building a cabin there. They had a busy winter running traps the whole time, and it was a good season for them. By spring, they had accumulated an outstanding catch—approximately 550 white foxes, seventy wolves and twelve wolverines. A fine bunch of fur.

Before leaving for their camp in the fall, they had made arrangements for a plane to come out in the spring. The plane, a Norseman, arrived to pick them up, but there was too much for one load. The capacity was something like 1,650 or 1,700 pounds. The bales of fur were quite bulky and their gear was heavy. Instead of leaving some of this behind, those danged guys shot their dogs, both teams, just so they wouldn't have to make another trip out.

That really made my skin crawl and no way in Christ's world would I so much have thought of it. If there was enough around to eat, I'd have said to hell with that racket. I would have told the pilot to take the fur in and return for a second trip. Those two guys had a world of fur worth a considerable sum of money. Maybe the extra charter would've cost another $300 or so; they had enough fur to afford it. Those poor dogs had worked their hearts out all winter long to make it possible for them to get that big catch. And the only return they got was a bullet in the head. When money gets in the way, a man's heart sometimes drops into his wallet. Even if I had no fur at all, I couldn't have done that.

Fate was to step in the next year when Stark returned to trap around Lake Providence. De'Steffany had decided against going back with him because wood had been hard to find and he was afraid there wouldn't be enough to last another winter. Going out to the Barrens without fuel was just plain suicide.

Stark flew out in the fall with a plan to return the following spring by dog team. If he hadn't returned by a pre-arranged date, he had asked the authorities to come looking for him. This was the last anyone ever saw of Jack Stark. When he didn't show up by the target day, a plane was sent to his camp landing near the cabin. Walking up to make the investigation, the first thing that captured the authorities' attention were the dogs, dead on their chains. They had starved to death and Stark was nowhere in sight.

After giving the place a thorough search, they found his gun leaning against the leg of a stage used for a food cache. In the chamber was a spent cartridge. It was determined that Stark had probably shot a wolf out on the ice near a falls close by and fallen through while trying to retrieve it. Neither Stark nor a dead wolf were found. De'Steffany really took it hard, and chartered another flight out to the camp. But this search turned up the same results as the previous one had. At last, the case of Jack Stark was closed with no clues to determine his fate.

As stated earlier, my oldest brother was tough on dogs, too. They worked for him, or else. One time we were together out by the Thelon River with our dogs. We were on a real rough trip over rough ground and had probably hurried too much. His team included two fat castrated dogs, and one of them died before we made it to our camp. Neither of them was in any shape for hard work because they were just too fat.

A short while after that, we went together on a trip from Lynx Lake to Granite Falls—sixty miles by dog team. The remaining castrated dog still wasn't pulling enough to suit Hughie. He said the dog was going to be destroyed if it didn't start working. For the dog's sake, I asked for the opportunity to drive him in my team and maybe straighten him out. Hughie agreed to give it a try.

The next morning when I came out of the tent, a pitiful sight greeted my eyes. The dog was lying dead near the sleigh. Hughie had enough, and hit the poor animal on the head with his ax. Jesus Christ, that hurt me, because there was a possibility the dog would've been all right if he'd just been given some time to get in shape.

Another time, he broke in a team the hard way, all blacks. This was a fun litter whose mother had been in harness before. She was put on the lead and they took off. I want to tell you, Hughie was taken on some circles that day before finally getting those pups broke in. They were all green, having never so much as seen a harness before that day.

At the time, we were out at our camp in the Barrens. George Magrum and I went ahead toward his place with Hughie hooking up to follow behind us. He was going to drive that team twenty-five miles on its first trip ever. This was in November, before the trapping started.

He tried to follow our trail, but part of the time went a good quarter mile either way from it. He was a dog driver and a rough one. They had to get the job done or, by the Jesus, he'd kill them. Sure enough, a couple of hours after we got in, here he came.

Perhaps the only reason that the female leader took them on was because she was afraid of Hughie. I was upset, and told Hughie he had gone too darned far this time. Those dogs had been tied up all fall and had no stamina. But I couldn't argue too long with success because Hughie had done his job.

He broke that team which had five pups in it, and raised them himself, driving them all winter. There has never been a prettier one. I sure felt sorry the next summer because distemper infected them and most died. The survivors were sold because, once a dog got distemper, he was never the same afterwards.

The way Hughie broke that team was not the usual method. Normally, time and patience dictate a better training technique. By the time a pup is eight or nine months old, he's big enough to go into the harness.

Seldom did I use more than one pup at a time. I'd put him into the harness as part of a team with the old dogs. There wasn't much to worry about, because the dogs were bred for this type of work, but once a leader took off with a fresh pup, they went like hell. I either got on the toboggan in a hurry or was dragged for a ways. At first, I would put mustard on the pup's traces and in his mouth to keep him from chewing the leather. Dogs don't like mustard, and one taste is usually enough to discourage biting of the traces.

Dog teams weren't restricted to males exclusively. Females made excellent workers, too, and I've had female lead dogs from time to time who were smart and hard working. But they required some special treatment. You had to wrap a blanket around them and make sure their teats were covered. If not, the exposure would freeze them.

Like everywhere else, the worst thing to get into our dogs here in the North was distemper. Once a dog got the disease, it really wrecked him. When the game warden came around, he could inoculate them with a serum to ward it off, but if the bug hit first, it was all over.

One year when white fox fur was very cheap, I trapped at the head of the Snowdrift River. There was no reason for going to the Barrens that season, so I remained in the timber to try my hand at catching colored fur. (The year was productive, too, because I caught forty-two wolves, a few mink, some colored foxes, three wolverines, and three otters.)

I was driving an excellent lead dog at the time, a real going hound. There was no reason to worry about breaking trail, because he'd bloody well do it or burst trying. Like most dogs during the trapping season, he was in good shape. Their diet consisted mostly of fat wolf carcasses which they liked and did exceptionally well on.

But surprisingly enough, this seemingly healthy and robust leader came down with distemper. Of all places, it happened at my second out-camp. It hit him so hard that I didn't dare bring him back to Fort Reliance. Exposing all the dogs there to the disease would've been very unwise. His eyes, mouth, and ears were badly infected and matted. Placing some fat wolf carcasses where he could reach them and tying him to a chain, I returned to Fort Reliance. Knowing my leader would be dead when I got back to this camp, it was a sad trip in.

Hughie gave me holy hell when I reached home. Finally, I convinced him that either way things went, they would work out. Should he die, the others at least would be spared exposure; if he lived, there would be plenty for him to eat.

The next time I made this out-camp, he was there curled up on his chain. Confident he was dead, I paid little attention while driving up. Then all at once, he jumped up into the air three feet high. Damn was I happy! Not only was that old bugger very much alive, but he was fat as a hog, even though the temperature had been fifty below for a week.

Something else that happened to dogs in the North was rabies. The only case of rabies I ever saw with my own eyes happened one year in the fall just before I was ready to depart for the Barrens. The dogs were raising a din about something and I went outside to find the cause. Down by the lake shore, the police had dropped off a sack of fish. Looking in the direction of the action, I saw a black animal I thought was a dog. But the thing, with its ears laid back and its tail drawn up against its belly between its legs, didn't look right. I was watching the danged thing trying to get the fish, when it dawned on me that it was a wolf.

In my haste to get outside to find out about the rumpus, I was still in my underwear. Of course, I had no gun. The wolf came toward me, and before I could get squared away, it ran right past me. It could've grabbed me easily, but instead went beyond to grab one of the dogs by the neck and shake him a few times. It let go right away, and then went over to another dog to do the same thing. By that time, I had come to my senses and reached inside the door for the rifle. I shot the wolf through the back.

After I examined it closely, I found it to be a female and a sick looking thing, at that. Skinning it out, I found tooth marks all over the rear end of the carcass. Apparently, other wolves in the pack had driven her off after she started foaming at the mouth and fighting them.

In my mind there was little doubt that the wolf had rabies. Proof came about a week or so later when the first bitten dog died. The second dog came out of it all right, so I probably shot the wolf before it had bitten through the dog's skin. I was very lucky to escape being bitten myself; it had run right by my leg to get to the dogs.

Thank God, most of the time well-treated dogs are healthy and filled with a vitality for life. They are amazing and, if a man trusts them, they will save him a lot of misery—sometimes his life, too. Many times, men take their dogs for granted, but when a man who knows dogs stops to think about their abilities, he finds their true value and worth.

One year, Phil and I were flown close to the mouth of the Hanbury River to find some new trapping territory. This area, called "no man's land," was adjacent to the Thelon Game Sanctuary. There was no natural boundary to mark the refuge and we stayed far enough away to avoid trespassing.

Soon after the trapping season had started, something went wrong with Phil's leg. He complained about it often, and I could tell he was in pain. He had suffered for as long as he could until, unable to stand it anymore, he had decided to return to Fort Reliance. Both our lines were out at this point, and fur was coming in, especially white foxes, which were particularly plentiful. Given that, I had tried to get him to stick it out until April. But the pain became more intense with each day and he'd had to give up. I knew nothing about his line—I hadn't been over it at all. Anyway, I told him to leave his lead dog and that I'd run his traps for him.

Phil had a policy of training two lead dogs and switching them around from time to time. We were forever looking for leadership qualities in our young dogs. The first thing we looked for was intelligence; a willingness to work was a close second. When we identified one with the tools to lead, we put him in the number-two position, right behind the old leader. As the old leader got older, he began to tire easily and react more slowly to commands.

The future leader running in the second spot learned these commands, because he was smart. Eventually, he would begin pulling to commands sooner than the leader, indicating a readiness to take over. Phil was always looking for this—he was very observant, too—and this young leader at the time was ready to take over.

When Phil left our camp, I traveled two days southwest with him. He knew the country quite well, having run a long line in that area the year before. But since he was hurting, I wanted to make sure there was no trouble—at least until he got close to the trees. We decided to begin the long drive toward Fort Reliance along the route of his old line. Several tent poles with a half-cariole of firewood were cached by one of his out-camps from the year before. Whenever a trap had been pulled along the trap line, the stake, or toggle, was left standing by a rock. Most trappers did this in case the same area was trapped the following year, thereby ensuring a trap anchor, however dry it might be.

As we were hooking the dogs to the sleigh before leaving our camp, I glanced toward the tent. Smoke was billowing from it and I felt sure it was on fire. Running back to it and going inside, I soon found the source of the trouble: I had done a very foolish thing and hung my caribou skin parka too close to the stove (an oil drum laid on its side with a hinged door and a stove pipe that traveled up to a dirt - and moss-insulated opening). Anyway, a huge hole was smouldering in the front of the parka. By the Jesus, was I mad!

I took it outside and beat the fire out. Then I pulled the hole together and sewed it with a needle and thread. It wasn't a very good job, but we were in a hurry to strike out and this was the best that could be done. After I finished, we immediately hit the trail. We traveled two or three hours before reaching a lake with a pack of wolves on our trail. We saw and heard them often, but couldn't get a shot at one. With ten miles still to go before we were going to reach Phil's old camp, and with the wind whistling through the hole in my parka, I was cursing my stupidity—and getting very cold.

Phil was using his new leader, but as we got close to the old line, he took him off and exchanged him with the old guy. The wind was blowing hard and the weather was becoming dirty, which caused me to worry about finding Phil's old camp. My concerns proved unfounded, though, because hardly more than a mile had gone by when the old white leader suddenly went sharply to the left. There had been no command, either.

And then all at once, that old bugger came to a halt. Searching around carefully, I found one of the old trap toggles leaning against a rock that stuck out above the snow. This was where Phil had set a trap the year before, and that dog had remembered. No sir, this wasn't the work of some dumb animal.

Phil marched him on, but it was an unsteady trip, as the dog stopped beside every one of the old sets. He was following the old trail, weaving around all the rocks to stay on it. We didn't need any map or compass; that dog knew the way.

I didn't need to worry about finding the camp site on our maiden voyage out, even though the dog hadn't been there for a year. Having gone on for a ways, he stopped—and there were the poles needed for our tent. The firewood was still there, too, and in good shape. By then, I was so cold and pleased that I could've given the dog my supper.

That was the way our dogs were and the degree to which we depended on them. If a man had a good leader who had been to a place before, he trusted him. The dog took him where he had to go. Another thing, after returning to the main camp with the white leader, I was able to run Phil's line because he knew where the traps were. He had to, because I didn't know the location of a single one.

That poor old leader died in a funny way, too. It was a year or two later, and Phil and I had returned to the Barrens early, as usual. We could never get untracked, even though we worked harder than normal. The caribou had missed us, and we couldn't get enough meat to last the winter. I wanted to stay and try to get enough fish to sustain us, but Phil would have none of that. There was little choice, it seemed, but to return to Fort Reliance (though if the same thing were to happen again, I'd stay and fish—to hell with coming back).

And we could've had all the meat necessary if we'd had a mind to. A herd of twenty muskox dared us to shoot them, but we simply couldn't do it. I couldn't have slept or eaten the meat knowing, even at that time, that they were an endangered species protected by law. There are some things even a man who lives off the land cannot do. Still, that sure was tempting. Especially when a man needs meat just to stay in a place for the winter. Those beautiful devils just stood there and, undoubtedly, I could've slaughtered the whole herd before they got away.

So it was back to Fort Reliance by way of the Lockhart River with the trusty old dog on the lead. This was still in November and, by the time we got in, we knew we'd had a trip. It was a son of a gun. We stayed around Fort Reliance until well after Christmas when we decided to take a chance and return to the Barrens. Hauling a big load of fish apiece, we struck out again on January sixth.

By the time we reached our main camp, our fish supply was nearly exhausted. Still, there was reason not to worry because we'd seen fresh caribou trails and so felt we'd eventually run into them. Their return was cause for optimism, so we put our traps to work. Fur was again plentiful and, soon, white foxes began finding our sets. There were unusually fat, and provided enough meat for the dogs to get by on, even though the fox carcass is very small.

In the meantime, we hunted for caribou. One day, we reached a point where Phil said we'd either find caribou within a mile or we could forget about them. Not going half that far, we came upon fresh tracks in the snow. The dogs could smell them and came alive with a start. Remaining behind to hold the dog teams, I told Phil to proceed and see what he could find. Pretty soon, I heard his rifle begin to talk. A lone, crippled bull ran past and I was able to finish him off. Phil returned, reporting the good news that five or six others were down.

We piled the meat up and cached it on the spot, and a feeling of relief swept over us. There were some traps on the sleigh—eighteen to be exact—that we set at various places around the meat. While traveling back and forth hauling the meat to camp, we caught nine white foxes for our efforts.

That night, the dogs were chained to their pegs in the usual manner and were given a big feed of caribou, their first of the season. The evening was a happy one. But the next morning when we went outside, the white leader didn't get up to greet us like the other dogs did. He was curled up, as if asleep. When Phil went over to investigate, he found a very dead dog. From all appearances, it seemed the old white guy had passed on in a natural way.

With the young leader still in Phil's team, our trapping wasn't halted or even slowed down. We hit the trap lines hard and, from January 20 to the end of the season, caught ninety foxes, twenty wolves and three wolverines. Upon returning to the Barrens, we knew the fox cycle was up because there were droppings on nearly every drift. And with the return of the caribou, came the wolves.

In the spring after returning to Fort Reliance, Phil mentioned to a Mountie at the barracks that his white leader had met his end in a quiet way. Right away the Mountie related a similar circumstance concerning one of his dogs. Puzzled by the unexpected death of this young and healthy dog, he had cut him open and found his digestive tract completely plugged with caribou hair. Unable to pass the ball of hair along, the dog had died. Right away, we saw the similarity between this and the white leader, and determined that the two of them had died from the same thing.

CHAPTER XI

Several dog names have remained constant in my teams down through the years. Whenever a dog died, a member of the next litter acquired his name. As a result, names such as Sport, Red Brandy, White Brandy, and Bucko were well represented in the teams I drove.

"Mari gon" was always a favorite of mine and one of the oldest. An Indian word meaning wolf, it was pronounced Morgan by me. And Kaviook, a name given to females, has also been used consistently since the early days.

I drove various breeds of dogs, but had primarily Huskies with a few Malemutes mixed in. But the last strain, the one I have now, comes directly from the Eskimos and is the real thing. This is the same bloodline they've used for years. Mari gan and his brother, Fritz, came from Tuktoyaktuk. Full brothers, I bought them when they were only a month old. Little Kaviook came from Sachs Harbor on Banks Island last fall and is part wolf. That's why she plays so darned rough. She comes into heat in March, and bred younger than any dog I've ever owned. Apparently, that comes from the wolf in her.

Blessed with instinctive powers somehow not granted to man, dogs have been closely observed by those of us who are dependent upon their abilities. I had to learn to lay my trust and reliance with these dogs—and sometimes this learning came the hard way, if the actions of my dogs went unheeded.

Out in the Barrens, the dog was the only weather man a trapper ever needed. During the dead of winter the sun rose around nine o'clock and set about three. With so little daylight, I had to be up and ready to go at first light in order to make the next camp before dark. The single obstacle preventing a day on the trail was the weather, which I watched like a hawk.

Just before dawn when first going outside, I could almost predict what the weather was going to be by looking at the chained dogs. If they were eager, and jumped up to enthusiastically greet me, I felt safe and looked forward to a day fit for traveling. But if they stayed curled up sleeping and ignored me, there was a bloody storm on the move and a dirty day on the way. The dogs were neither lazy nor fooling around about this. They didn't like being tied up with those chains, and preferred to be in the harness moving along the trap line.

There were times when I ignored these clues, figuring it more important to get to my next camp. I'd hook them up and strike out anyway, but usually ended up turning back before getting very far. A few times I kept going and got caught in a big storm. They proved to be very smart animals, and I had to learn to trust them, because they were usually right.

A good team will cover a lot of ground, too. On long trips with a heavy load, they can travel four or five miles per hour without tiring. On short hauls when traveling light and in a hurry, they'll march along at a clip of seven or eight miles per hour. When in good trail condition, they'll do it easily, too.

Whenever I could, which was most of the time, I rode on the toboggan. That's one of the reasons that I raised dogs. I didn't have to walk except in deep snow where the dogs broke through. When this happened, I had to go ahead on snowshoes to break the trail.

There was no set number of dogs in a team and different trappers used different counts. My teams were made up of either six or seven. In later years, Delphine and I would pile half a dozen stiff wolves on the sleigh, then climb on, too. I'd cluck to them and off we'd go. Those guys would poke right along and do it easily.

Since working on a full stomach was good for neither man nor beast, the dogs were fed at night, and so went to sleep with a satisfied appetite. This helped keep them warm during the long, cold nights, and made them ready and eager to go in the morning. If their bellies were stuffed just before hitting the trail, their enthusiasm for traveling wouldn't be so keen. They were fed small portions—the same amount each day. Again like humans, dogs seemed to do better when they weren't overfed. Following this policy of once-a-day feeding, I never had to use a whip.

The key to what dogs will eat comes when they are puppies. If they are started right, they'll learn to eat about anything. They should be trained this way because, in the Barrens, a man never knows when he might get stuck without dog feed. Mine wouldn't hesitate to eat what I had, although there was one meat I had trouble making them like: colored fox carcass. They wouldn't touch these unless they were on the verge of starvation. Unlike the white fox, which I've seen the natives eat, the colored fellow isn't fat.

A good lead dog has a homing instinct that ranks with the best of animals. He is like a saddle horse on the prairie or a wolf on the Barrens. I've seen a lone wolf make tracks all over the place while hunting the countryside. He would hunt this way and that way, but when the bugger was ready to go to his den or to the pups, he didn't back-track or waste any steps. Striking out on a straight bloody line, he'd get home in a hurry.

The wolf hadn't been watching for landmarks or paying any attention to where he was while hunting. He didn't get lost because his instinct told him how to get home. A dog is like that when he's in an area with which he's familiar. I learned to let mine lead the way because he knew where camp was. Good dogs are wonderful animals.

Most people on the outside are probably of the opinion that the lead dog is a big, tough bully who can whip the whole team in a fight. Seldom is this true. It seemed to me that the rest of the team often had it in for the leader. Whether through dislike or jealousy, they would try to nip him whenever the chance presented itself. The leader wasn't necessarily the biggest dog, but for intelligence and willingness to work, he was number one.

Everything wasn't always pleasant with dogs. Sometimes, they did things to aggravate a man and caused some real problems. A dog that became contrary could be a headache and cause a man to talk to himself.

There was once a fellow by the name of Walter Langstaff, an American, who trapped out of Fort Reliance and on into the Barrens. He went out for years and was a lone wolf. Also tough and a hardworking bugger,

Langstaff wouldn't let a blizzard slow him down—and the result was usually a fine catch on his part.

One winter, his dogs gave him quite a surprise. He was running his line and came up to a trap which had a wolf in it. Langstaff always got around early in order to reach the end of his line because his trap line stretched for a greater distance than most. This happened at daybreak just before sunrise.

Stopping the dogs a short distance from the trap, he went over to kill the wolf. After doing so, he turned to go back to the sleigh and found it gone. He looked ahead, and spotted a herd of caribou crossing the trail with his dog team in hot pursuit.

The team went out of sight over a rise. They got completely away and, no doubt, became tangled up somewhere. Perhaps they became involved in a scrap because they were fighting fools. Langstaff took off after them on foot. He circled and circled all that day and part of the next, but he never did locate them. When he told me about that, I wondered how they kept from chewing out of the harness. They were part Airedale and a fighting bloody outfit crazy for caribou.

This happened in March, which would've spelled doom for some men, but not for Walter Langstaff. He loaded his big freight canoe, either making a sled for it or using a toboggan, and shoved that danged thing for miles over the ice until hitting open water. He was headed for Fort Fitzgerald and, by God, he made it, too. He had that canoe loaded down with all his gear and bales of fur. That man was a trapper, usually having some colored fur and plenty of white foxes, but few wolves.

Helping him portage that thing one year, I know how heavy his canoe was because I suffered on one end of it. He used to drag it to his camp, with dogs making a kind of skid behind an old sled. He angled a couple of poles back from the laced top, then placed the canoe on the poles so one end was off the ground. The dogs took it across the portages that way, too. Langstaff was a man who did things with what was available to him, and losing the dogs didn't bother him.

We also had a dog problem and his name was Red Brandy. My brothers had him in their teams most of the time, but I drove him some. Two things he stood for were toughness and meanness, and he could go from good to bad in an instant. By our standards, he was one of the best working dogs ever, but he would fight anything at the drop of a hat.

One time, Hughie and I were approaching Granite Falls by canoe and had pulled to shore for the portage around it. The dogs were milling around and, as usual, Red Brandy got into a fight with two other dogs. It was two on one from the start because Red Brandy was hated by all the dogs. Not only did he like to fight, but he was a tough bugger with it.

The fight broke out on the river bank, but soon progressed to the river, which was quite swift at that point. Pretty soon, they got far enough out that the current caught them. The fight stopped, and the two dogs got squared away. They swam to shore, but old Red Brandy wasn't quick enough, and he never made it. He was just too upset to realize his trouble, and over the falls he went. I told Hughie he didn't have to worry about his damned fighter any more.

Granite Falls is a cascade of about a quarter mile and there was no way for the dog to come out alive, or so we thought. With nothing to be done about it, we went ahead with the portage. Arriving at the end of the portage below the falls, we looked across the river. I'll be damned if there wasn't old Red Brandy shaking the water from his coat. We were relieved to see the bugger and I think he was happy to see us, too. I don't know how he survived the ride, but nothing serious seemed to be the matter with him.

Through the years, a considerable amount of caribou met their death in just this way. They hit a straight line during migration and hold it so faithfully, they'll seldom go around anything except a burned area. They swim across lakes and rivers as if they weren't there. Whenever they were unlucky enough to cross at places above falls such as Granite Falls or Parry Falls on the Lockhart, some got caught in the current and were swept over the edge. A caribou is a powerful swimmer, but once his nose goes under, it takes very little to do him in. Maybe he panics and gets his head under when he hears the falls. In any event, several dead ones were found below these falls every year.

I cannot think of Granite Falls without one memory standing out. This darned cascade had to be detoured either by portage early in the fall or by taking the long way around with the dogs during the winter. I got tired of this and, one winter, approached the bloody thing with an impulse to drive the dog team over it. It was a damned silly thing to do, but I went over anyway. Talk about one hell of a ride! Under my feet, the water rushed and gurgled. Those dogs heard it and danged sure never slowed down. When I told Hughie about this later, he guaranteed me he wouldn't try that for all the tea in China. Of course, he was right to feel such apprehension; I never had nerve to try the stunt again.

One year when Phil and I were together, I was driving Red Brandy in my team. Traveling together, we saw some ravens circling over the remains of a wolf kill. When we got close enough, we could tell the wolves were still there. Phil went ahead on foot with his gun, hoping to get a shot at them.

I stayed behind trying to keep the two dog teams apart. But as hard as I tried, I couldn't do it. That danged Red Brandy was the worst of the lot, and wouldn't quit until starting a rumpus. He got the job done with a big fight that saw both teams going at it.

Hearing all the commotion, Phil came back and helped to break things up. Something had to be done with Red Brandy, so the next day we took him down and snipped the two top fangs off with a pair of pliers. That guy was nothing but trouble, and something had to change his attitude if he was to remain a working dog in my team, but I hated to do it. After that, he still had some fights, but soon learned to be a diplomat because he was unable to hang onto his opponent. He was mean, but also became one heck of a beaver dog. If Phil shot one, he would swim out in the lake to fetch it back.

A few years later, I had another bad fighting dog. He was simply too full of energy, and that got him into trouble. One of the best pulling dogs I ever had, his name was Beauty. Every morning, he was the last dog to be put into the harness because he couldn't stand still. He either started a fight or ran off with the toboggan. When I put him in line, I always grabbed the toboggan because he could really take off.

I could count on one thing with that dog every day. He would bark every step of the way for the first hour on the trail. This made me very angry with him, but he remained one of my favorites. At night, he was also the first one out of the harness. He was an ornery son of a gun, but what a dog!

Cold weather is one thing not to be concerned about when it comes to a northern dog. They can take all it has to offer, and sleep right out in the bloody open, even during three or four-day Barrens blizzards, without suffering any ill effects from it. On our first year on the tundra, we built some little kennels for the dogs. But it was a waste of time because they wouldn't even go inside them, preferring to sleep outside on the south side for protection from the wind. Apparently, it was too hot for them inside the dog house.

Every morning in the winter, the dogs were a sight to behold all curled up in one furry ball, fast asleep. They were warm and comfortable, and though I'd sometimes see an ear prick up to catch some small sound, not another muscle was moved. If the sound proved insignificant, that ear would drop like a periscope. The nose was buried deeply in the base of the tail.

Foxes and wolves sleep the same way, with the hair around their nosewarmer becoming rough and tangled. When fixing the fur for the market in the spring this had to be combed out and straightened. All animals had to protect themselves from the cold by covering every bare area so nothing would be left exposed.

A considerable amount of snow fell on the tundra, and was continually blown around, but there was no danger of the dogs becoming snowed over and suffocating. They were too intelligent to let this happen. There was an instance one year at Beaverhill Lake between Christmas and New Years Day when we were hit by one of the worst blizzards I'd ever seen. It came at a very unfortunate time, because there was a big white fox run on.

The first night the dogs began howling miserably, so I went out to check them. They were up, walking around stiff-legged. Upon closer inspection, I realized that their legs were freezing. They were stiff like a board, so Delphine and I brought them into the cabin and went to work to relieve their pain. They were very lame and their legs were nearly useless.

We were at the cabin that Red Noyse had helped build a few years earlier. It had uprights for supporting the logs so we tied the dogs to these poles and kept them inside all that night. They never made a mess on the floor or anything, and I've never seen dogs so content in my life.

That particular blizzard was one of those dirty ones driven by sand and sleet. The strong wind had parted the hair on the dogs, and that had allowed the freezing air to get to their skin. This kind of storm was a dreaded thing on the tundra. Indeed, this one was so bad that I had been afraid to build a fire in the cabin. A downdraft meant the possibility of catching the cabin on fire was too great. And once that happened in that type of storm, there would've been nothing we could do about it. We crawled into our blankets to stay warm, and let it blow. We put the dogs back outside on the second day, and the storm eased up on the fourth.

Another trapper, Fred Riddle, was located some thirty-five miles to the east and a little south on Mosquito Lake. That same storm caused him some temporary concern. He had constructed a regular kennel of little cabins for his dogs. Riddle was quite a man. Along with being an outstanding trapper, this was a man who treated his dogs right. I doubt if he ever hit a dog and, when giving commands to his team, he hardly more than whispered. They always heard him, and a more disciplined team didn't exist.

The morning after this storm hit, he looked out and discovered his dog houses to be snowed under—completely covered by blowing snow. At first glance, he couldn't even tell where they were. This scared Riddle, because he thought all his dogs had suffocated.

Going outside, he found the little mounds of snow that indicated the location of the dog houses. He found the worst one, and started shoveling the snow away with not a sound coming from below. After he'd scooped the opening big enough, his dog burst out like a shot out of hell. All were eventually dug out and found to be in good shape. Somehow, enough air had been trapped down there to keep them alive. Good sled dogs are tough, and seem to come out of things like a miracle.

While on predator control for the government and using poison to control the wolves, I had to be careful all the time. That stuff was deadly for dogs, too. Delphine and I were forever aware of the danger it presented, and always worried about any possible misuse.

At the time, I was driving an excellent dog, Bucko, in my team. He was caribou-crazy, and usually spotted them well before the other dogs or I did. When he did, he'd try to pull the toboggan in their direction, wanting to take up the chase.

One day, that son of a gun somehow got a shot of the poison. Delphine and I were completely sick at heart, and fully expected the worst. There were some dead wolves on the sleigh as we had just returned from the line. All the time we were pulling them in, the dogs never touched them—I had made absolutely sure of that. When I thought about it later, the only answer I could come up with was that maybe a little frozen saliva had fallen from a wolf's mouth and Bucko had licked it up. He all at once keeled over, and severe spasms took over his body. Disgusted with my lack of foresight and sorry for my dog, I went over to where he lay, and forced his locked jaws open. He was a very pitiful sight and scared to death.

I asked Delphine to go quickly into the cabin and mix some mustard and tea. She did this rapidly, heating the mixture and bringing it to the stricken dog. I forced all of it down him. Pretty soon, that bugger rose to his feet and vomited the whole mess up. After making a jump or two, he once again keeled over. Again, Delphine returned to the cabin, this time cooking some rolled oats which I also forced him to eat. He didn't want any part of it, but I pushed it down him anyway and rubbed his throat to make him swallow. He was a very weak dog, and I knew he was going to die. I tied him to his chain, and left him there for the night.

The following morning when I went out to look at him, I received quite a surprise. He was far from dead and, seeing me, that brute leaped into the air like a puppy. I've never been quite so happy about anything. That piece of business taught me a thing or two about poison, and I never repeated the mistake.

There are certain things dogs didn't like to do and traveling on ice was one of them. The average leader wouldn't go out on bare ice because he couldn't stay on his feet to generate the power he needed to pull a load. He walked all spread out, completely helpless and unable to pull. And a few miles on that kind of surface tore his feet badly. One of my best leaders refused to go on glare ice. Whenever we were on a lake with patches of snow around, he would zigzag around, so he'd miss all the icy spots.

Fortunately, the North had very little of this. Soon after freeze-up, the snows arrived, and rarely more than a few days behind. With snow on the ice, this problem was a minor one. As a rule, most of the ice travel came in the spring after the snow melted, but the ice wasn't slick like in the fall.

Perhaps the most important thing a dog has to learn is to stop when a caribou is close enough to shoot. This is an important lesson, but easily taught. It doesn't take him long to catch on, but he has to understand. Normally, a caribou lets you know when he is going to stop because he will suddenly slow down. He may be running all out, but if he doesn't catch a scent, he will stop. A caribou is a nosey cuss and comes to a halt to figure out what you are.

There was a head line from the driver leading all the way to the leader. It went through all the rings on the right side up to the leader where it was tied—not to his collar, but around his neck. A sudden pull on that followed by two or three turns around the lazy back and a bull couldn't go anywhere. The dogs would be bunched up with nowhere to go. By then, the rifle would be out and ready, with a short step and a drop to one knee completing the act. Man and dogs had to act as a team in such cases, as timing was important. A well-trained team of dogs would stand still right there until the shooting stopped.

With a very keen sense of intelligence, the dogs came to learn the importance of caribou hunting, determining what a man wanted and when to stop. Some of them bark, and it's obvious what that could do to a caribou hunt. I was always very lucky in this respect, because barking aggravated me and I soon put a stop to it.

My dogs were very good until seeing a caribou drop. Even though they couldn't go anywhere, they would give a lunge. I thought it best for them to go right up to the downed animal, grab a mouthful of his hide, and shake him. The dogs then figured they had a hand in bringing down the caribou, so I let them bite and growl for a while. Pretty soon, I'd yell a command and we would go off after the rest of them. The team got used to this, and I usually had good caribou dogs because of it.

In this country, a good dog is just as important as warm clothing. Evans Peterson could attest to that. He and his brother, Martin, were trapping in the country before we ever came out. They were around Beaverhill Lake for a while, and came in every spring with a good bunch of fur. Most of the time, they took this fur to Edmonton and sold it at the auction. This one year, Martin went from the auction on to Vancouver where his family was located.

Intending to return to the Barrens with Evans in the fall, he got blood poisoning during the summer and had to remain behind. The risk was too great, so Evans was told to go on alone. Martin's line had been the most productive and Evans wanted to use it. Even though this sounded good, there was a problem. This was new territory and Evans didn't know the country.

Not knowing the location of the traps or out-camps should have been enough to change his mind, but Martin told him to stop in Snowdrift and pick up his lead dog on the way in, which he did, along with his own, and then struck out for Beaverhill Lake. He drove his leader all the way, letting the dog find everything for him. He arrived at the main camp in good shape.

After resting up, Martin's leader was put to the test. They struck out in the direction his brother had always taken when running his line the year before. Immediately, the dog took it up, and he seemingly made every turn right in stride. Evans let the dog have his head, and made no commands. He kept saying sweet things and treating the dog, who was happy as a kid, really well. They pulled up to the first out-camp that night and, the next day, they made the second out-camp. Not only did they find these camps, but the dog stopped beside every one of the old sets. Evans had never been there before, but the dog took him to everything he needed to find.

CHAPTER XII

Even though most of my years have been spent in the woods and beyond, I've known that the opinions and beliefs of men come from one of two basic exposures to learning: personal experience and research from reading. The two are probably combined often, but I doubt if one ever achieves a depth of knowledge in an area away from his specialty. If the subject in question is one concerning nature, for example, and a particular man gains his knowledge first from the written word, before trying to get insight from personal experience, he naturally draws his conclusions from the words he has read.

On the other hand, a man like me who has learned things from firsthand experience in a natural setting as part of his life's work, will lean toward what he has picked up this way and finds it difficult at times to accept whatever he reads that is contrary to his beliefs.

This distinction in the acquisition of knowledge has led to difficulties and confrontations between men who have taken a lot of time to learn the same subject—but from different approaches. Also, it seems like pooling this information is usually not acceptable to either party, because a middle-of-the-road attitude is normally adopted by a third party who knows nothing about the subject. Neither the practical man nor the book learner is happy in this case, because doing anything short of what each believes is a failing solution to the problem.

Specifically, I have in mind the battle over the wolves which rages on in the North with the learned-by-experience men—including trappers like me—arguing our side, and the scientists and humane societies arguing theirs. I personally know wolves and their habits and consider them to be enemies because of the destruction they do to the livestock of the North: the caribou.

I maintain this belief very strongly and, while no doubt many will hold it against me, I think there are good reasons for it. I feel that when anyone has an opinion on any subject and can substantiate it with sound reasoning and first-hand knowledge, he is not a man to be scorned.

Seldom does anyone interview any of us who have lived with the wolves. There should be something of value to be told after knowing and pursuing them in their natural environment for over fifty years.

I don't hate wolves. On the contrary, I love the animals. They have always been a very important part of my life, native and plentiful in the area where I've lived and worked. Trapping them has helped my basic existence as they have provided a great deal of money for me. Nevertheless, I still have to believe the only good wolf is a dead one. I'm convinced beyond doubt that they cause much more harm than they do good. In this case, balance of nature be damned.

There need not be any fear out there because the wolf will never be exterminated as he is just too sly. That is, unless all the caribou die off. There is just one hell of a lot of country out there in the Barrens with not a soul in it, so there is no way to ever get them all. Crowded conditions caused by too many people will never bother them because a man can't very well live on the tundra unless he's a trapper.

All wild animals are the most interesting and cutest things in all of God's world, especially their young. Quite frankly, wolves are no different, and I could observe and study them forever without tiring. There is no person alive who couldn't amuse himself for hours watching a litter of wolf pups frolicking around their den. But as pleasant an experience as it might be, most people enjoying it fail to understand the true character those playful pups will someday assume.

I'm in agreement with the experts that wolves will, on occasion, kill a sick or weak caribou and eat it. Everyone knows that is true. But it seems as though, without gaining all the available facts, it's always some chicken-hearted man or woman who takes this part and, in their writing about them, plays up how cute and harmless they are, and how enjoyable they are to observe. It's so darned hard to reason with people like that who just aren't able to get out and see what really goes on. And I am in full sympathy with their side of the story, too. But if only they would go to the trouble of seeing the devil in the critter.

Perhaps I'm just a bull-headed old-timer who has spent too much time living out in the wilds with wolves. But I do know what they do to the caribou herds by spending all that time in the Barrens and witnessing it all. Those wolves are with the herds twelve months a year, and will make a kill whenever hunger prompts them. The only natural living enemy of the caribou is the greatest predator of them all: the wolf. Their reduced numbers have to at least in part be directly attributed to the wolves who prey on them.

Of course, old Mother Nature doesn't make it an easy and automatic chore for the wolf to hunt down his caribou. He has to work for his meat, and encounters the occasional problem in getting it. Yet I've seen a hundred times more fat wolves than poor ones, and have never seen one starve to death. The whole thing comes down to one simple fact: if a wolf gets hungry and caribou are around, he will get one—sick, weak, or healthy. And that's a bloody fact of life on the Barrens. No, most of those preservationists won't argue with us old-timers. I suppose we would just be too far apart in our views to get along.

Wolves are of the most intelligent animals in all the wilds of the world. One thing they won't do is charge into a herd of caribou intending to kill the first one available. They will cleverly cut a single animal out of the herd and get him off to the side by himself. I've witnessed this scene countless times, with a set of caribou tracks leading away from the main herd. From time to time, his tracks indicate where he tried to return to the herd, but there would be two or three sets of wolf tracks on each side of him. After the wolves cut him out, they keep the pressure on, and lead the doomed caribou down a trail over which he would never return. The wolves herd him away and chase him to a slaughter he couldn't avoid. Most times, I saw positive proof of the kill in the circling ravens who waited from above to reduce the caribou to a few scattered bones. When caribou were around, this scene was duplicated every day. There have been times when I've found as many as twenty kills in one day.

It's almost impossible to convince me that these new-fangled ideas about saving the wolf are right. They should neither be completely exterminated nor allowed to freely flourish. The authorities who make the laws must make up their minds on one thing: do they want a country full of wolves or a country full of caribou? Both can exist, but the caribou herds would be better off with fewer wolves preying on them.

I've seen it lots of times. There would be a big territory completely void of caribou and not a wolf track visible or a single howl heard. And then all at once, the caribou would come from nowhere and leave their tracks everywhere. In short order, wolf tracks, too, would begin to show themselves and howls would become common at night. When the big herds were around, I've heard as many as five packs howling all at once. That could be more than fifty wolves, all of them with big appetites waiting to be satisfied. That many wolves would probably kill more than twenty caribou a week. I know what I'm talking about and these are not exaggerations, but facts I've accumulated after spending all those years competing with and against wolves.

One of those days comes to mind that helps prove the point about the damage wolves do. At the time this surprised even me, and I knew wolves pretty well. One spring, I was passing through at the head of the Snowdrift River on my way back from the Barrens to Fort Reliance. For some reason, I was out of meat and needed a caribou badly. There were plenty of tracks, but it was a long time before I saw a fresh one. In my hunt for caribou and a crack at one, it became obvious to me that no one had trapped the area that winter because wolf tracks crisscrossed everywhere I looked. They were thick because there was nobody to control them.

After finding that fresh track and with nothing else to do, I began counting caribou kills, and found one and sometimes two on every bend. All of the kills were fresh, and I didn't go very far. There wasn't a piece of meat remaining on any of them and, many times, other packs had come along and chewed the bones, leaving nothing edible remaining, not even a head. It was characteristic of wolves to follow their own, especially when caribou were around. Anyway, this was just a single day when I counted twenty caribou killed by wolves.

I would've been satisfied to discover some bones intact just to get the marrow, but the only place I could find any was over the river bank. A cow and her calf had been killed above and had rolled down the bank as the wolves pulled at the meat. Luckily, I looked down and was able to recover a few bones.

Once, there was a fresh set of caribou tracks made only that morning, but fresh wolf prints alongside relayed all-too-clear a message as they pursued him. He was a stray heading south and I'm sure the bastards got him. In this area of deep snow they would tire him out and hamstring him. I thought about going after them, but the scarcity of meat prompted me to continue to Fort Reliance.

For a long time now, I've watched them work on caribou with one conviction consistently coming to the front: If there was a fat caribou around, they wouldn't bother a sick or weak one. They prefer a healthy one with a good layer of fat and can tell the difference. Hell, they are smart, and knowing the difference is their business.

When the horns of the big bulls are in the velvet, they become easy prey with the wolves especially tough on them at that time. Since his new rack is heavy and cumbersome, he becomes easier to bring down. Protecting himself with his antlers is almost impossible, because the velvet is quite soft and filled with blood vessels. In fact, they are almost useless, as they can be easily cut through with a good hunting knife. Wolves really play hell with them at this time.

The time at Backs Lake when I got lost and went without sleep for nine nights taught me something else about wolves. They certainly don't have to chase a caribou to kill him. That masterful piece of work was done by a single wolf during a mild blizzard. If a loner can sneak up on a sleeping caribou and kill it before it can get up, think what a pack is capable of doing. Yes, that was the same cute little pup who used to play outside the den with all the innocence of the young. But now, he was a killer turned loose on the caribou, and having no trouble getting his share.

And another thing. In the early days, the wolf helped us on our way back south in the spring from the Barrens. Seldom carrying any meat on our fully loaded sleighs, we completely depended on wolf kills along the way to provide enough meat for dog food. We watched the sky in all directions in search of ravens circling over a kill. When we spotted a flock, we would make a bee-line for the kill. But all that changed because the wolves became too plentiful, cleaning up all the meat before we could get to the kills. There were just as many kills in later years, but the wolves were more numerous, following each other around to reduce each kill to a stack of bones.

A wolf bitch and her mate will go right out on the open tundra to summer with the caribou herds. There, they dig their dens and raise their pups, killing a caribou every two or three days for the family. If they can't drag part of the meat back, the old wolves will fill their stomachs and return to the den where they'll throw the meat up for the pups to eat. Later, when the pups get big enough to tear the meat off for themselves, they will drag a quarter back to them.

The wolves hunt all the time, because they won't use the entire kill. As soon as they take off for their den, the ravens and gulls go to work on the remains. Anything these two birds don't eat, they ruin by leaving droppings all over it. Wolves are particular about this, refusing to eat meat made dirty by bird manure.

A mated pair will remain together and, with the year's litter of pups, sometimes make up a pack that runs together all winter long. I came to know it happens this way because, while on predator control using poisoned baits, I sometimes got an entire pack at the same time. Inevitably, there would be an old male and female, and the rest were yearling pups. Occasionally, I'd get eight or ten in one bunch, all getting the poison at the same time.

Contrary to what the modern belief is, wolves can run any caribou down and kill it. They might work together in relays, and their added stamina gets the job done. This is especially true in deep snow, where the caribou gets too hot and are worn down. Once wolves close in, they tear big chunks of meat from the animal's flanks. This is kept up until he's hamstrung and too weak to resist. They then pull the caribou down and it's all over.

Wolves get pretty good at following a trap line, and can play hell with fur caught in a trap. If a fox is fat, they'll eat it; those in poor flesh, they simply tear to pieces. When it comes time for a wolf to eat, he wants fat meat.

Wolves don't hesitate to go after lemming, either. Not all wolves follow the caribou out to the Barrens in the spring. An occasional few stay behind in the timber, but they are very scarce. The ones who do were probably delayed by a female who was too close to whelping her pups to travel trying to keep up with the herds. With the pups on the way, a den was dug in preparation for the birth.

The lemming population would be hit hard when the caribou weren't around, because there wasn't anything else around for the wolves to eat. And those wolves really go after lemming in a big way, eating enough to fill up before returning to the den to regurgitate it for the pups. If they could get a hare, gopher, or caribou calf, they would carry it back to the den when the pups were big enough to handle bigger chunks. There are no better providers for their young than a pair of wolves.

Today, there's no doubt that something has to be done to control the wolf population. Before, the trappers helped this problem greatly but, except for Fred Riddle, there's not a soul out there to act as a buffer between them and the caribou anymore. The wolf population increases every year, and the caribou numbers fluctuate accordingly. The natives can't be expected to do anything to help because they don't trap anymore. To further complicate the matter, many of them are superstitious, and won't shoot a wolf. Even when one is caught in a trap, they turn it loose.

A native family named Drybones lives down on the point by me and the wife would get after wolves, though. Sometimes, she becomes a bit too eager. A year or two ago, she made a big mistake and, seeing this animal running across the ice in front of the tent, she grabbed her rifle and dropped it with one shot. When someone went out to bring it back, they discovered that old lady had plugged the pet dog from the weather station. A big collie, no less.

The sight of wolves on Charlton Bay was by no means an unusual one. One year not long ago, I was looking out toward Maufelly Point and spotted fifteen of them playing around. Delphine and I went after them right away and, before long, had eleven of the buggers ready for the Edmonton fur auction.

It wasn't necessary to use poison to get wolves, as I could catch them in traps with relative ease. One of the favorite methods used to entice them came about this way. When going out to the Barrens in the fall, I'd take some long poles along. At a place where wolves were known to cross when the caribou were around, I'd cut a hole in the ice as soon as it was safe and place the butt end of a pole in it. It would freeze there, with the other end sticking well up into the air.

Once the season opened and the fur became prime, the dog team was driven to make a trail as close to the pole as possible. Wolves followed dog trails all the time and weren't after them or me. They were either curious or wanted to walk where a trail had been broken for them. Anyway, I stopped the team by this pole and allowed them to urinate on it. In this respect, a wolf is just like a dog.

After the dogs had done their duty, we drove them a few feet away while setting traps around the pole. We made ice sets with the trap carefully covered with snow. If a wolf following that trail passed close to the pole, he couldn't resist taking a leak where the dog had. Plenty of wolves met their Waterloo in this fashion during the years I trapped the Barrens. When caribou stayed around all winter, they caused problems by springing the traps. They would smell the salt in the urine and paw around until they kicked the trap. As a matter of fact, I once caught a caribou in a trap at a set such as this.

Several trappers working on predator control were stretched across the Barrens, from Churchill to the Coppermine River. In conjunction with our regular trapping, we were paid extra to go after the wolves. There were seven of us that I knew about, and each was paid $240 a month to bait them. The government selected us primarily because we were the regulars who spent every winter at the edge of and beyond the tree line. All of us pretty well had established territories, stretching in a line across nearly 500 miles of country.

Something had to be used to attract wolves to the poison, so we kept the caribou paunches safely aside until the meat was cached. Later on, once the trapping season started, these paunches were filled with heavy blood scraped together from where the caribou were dressed. Strychnine pellets were then pushed into the blood and the whole thing was buried in the snow and covered with caribou skin. The deeper it was buried, the better it seemed to work. Wolves were crazy for blood, and could smell it from a great distance.

Strychnine wasn't something to be thrown around at random, so I was very careful with it. If caribou were killed during the winter, the fresh kill was baited, but none of the meat was wasted. The carcass was taken for food and the paunch was baited and buried. Wolves were attracted to the scene because of all the blood around, and the caribou hide over the paunch made him think something easy had been found.

I used to wonder if there was anything a wolf would refuse to eat through just to get to a baited paunch. I gave it a try. Coal oil stinks and when you get the stuff on your hands or clothing, it's almost impossible to remove. I suggested this test to a trapper friend, and he said a wolf wouldn't touch the bait. So we made a wager. Locating a place where wolves were certain to cross, I fixed a bait absolutely soaked with that stinking coal oil. When I returned to that place a few days later, I found two dead wolves beside the set. This convinced me a wolf will wade through hell to get a caribou.

A real old-timer, Fred Riddle is a superstar around these parts and there's no denying it. Also on predator control, he had some fantastic years when he came out with eye-popping catches. One year on Aylmer Lake, he hit it in a way that most trappers only dream about, and brought out over 400 wolves. These wolves weren't there by accident; over 200,000 caribou spent the winter in the area that year. With the help of two Eskimos from Bathurst, Riddle made that big tally and, right then, we dubbed him "King Wolf of the Barrens."

Riddle returned to the same place the very next year, but the caribou didn't, having wintered in a different area. He only caught something like eleven wolves that year, again proving the point that wolves are simply not around to be found unless there are caribou about.

Riddle has been at the edge of the tree line for a good many years now, and has hit the wolves nearly every year. There have been other years when he has also brought out 300 to 400 skins. That man was a super trapper, and has been at the game longer than any of us.

Even though I caught my share of wolves each year, I was never able to hit it big like Riddle. Usually well out in the Barrens, my only chance at getting wolves was in the fall, when they came out, and again in the spring, when they returned to the tundra. Since the caribou wintered to the south of me, that's where the wolves were. A man sitting in the trees with the caribou had easy pickings, because he had all season to work on the wolves.

Regardless of where a man worked on predator control, once a wolf got his poison, it didn't run off somewhere to die. Death came quickly, as evidenced by how closely to the bait the wolves fell. None were lost, even after a blizzard covered them over, because their tell-tale mounds were easily seen. We dug out the bodies, by now frozen stiff, with the snow shovel and loaded them on the sleigh.

In some ways, poison was a more humane way of taking wolves than traps, which held the animals firmly and resulted in a slow death or a long wait until the trapper came along to knock it on the head. In any case, we brought all wolves back to the camp to be placed either on a stage or the roof of the cabin until an opportunity presented itself to skin them. Normally, they weren't skinned until a day came along when traveling over the line was impossible. Out there, a man didn't have to wait very long for such an opportunity, because every few days a blizzard would come along to keep him in camp.

Once a storm came down, we would carry three or four wolf stiffs into the cabin. They were too rigid to do anything with, so we hung them over night until the following afternoon or night, when the legs were thawed well enough to skin. Since the cabin at Beaverhill Lake was quite big, we rigged a pulley with ropes across one end. Delphine and I had plenty of room to keep our fur with a drying rack well away from the heat.

Just thawing the legs was all that was required to start the skinning operation. That meant slitting the legs and skinning them. After that, we would lower the wolf to the floor for the rest of the skinning and raise another in its place.

With the legs done and the wolf on the floor, we made additional slits for the next step. After positioning the skin precisely, I would go to work on it with a hammer. Since most wolves were fat from eating caribou, the rest was easy. I pounded away with the hammer right where the hide and meat connect. The hide would peel away. Since it's so tough, I never worried if I didn't hit exactly where I wanted to; the hide didn't puncture easily and I very seldom made a hole.

The wolf wasn't allowed to hang long enough to thaw out completely because I couldn't have stayed inside with it. The dead wolf carcass would've gotten high quickly and the smell would've chased me out. Getting the job done while the body was frozen left no odor and the pelt came away from the carcass very easily. A wolf could be skinned in an hour.

We would keep the carcass with the meat used for dog food. The wolf was quartered the same as any animal and in the long run I considered the meat superior to caribou for dog food. Dogs seemed to eat it better, too. Poisoned wolves were also fed, but I was always hesitant about serving any part of the body cavity. Magrum, Riddle, and De Steffany didn't let it bother them; they used to gut their wolves, then slosh them out with water, and feed the whole thing.

Although to my knowledge none of their dogs was ever poisoned, this always appeared too risky to me. I wouldn't do it for love nor money, and used only the quarters and, in an occasional emergency, a little meat cut from the back. I discarded all parts of the head because the mouth was surely loaded with the poison. It stayed in the saliva.

Wolves have been known to become quite tame when getting accustomed to a man's presence. One hung around Martin Peterson for so long that it became known as his tame wolf. That was out beyond Whitefish Lake near what was called the ridge camp. He kept telling me about the thing hanging around his camp, but I could never see it. The wolf wasn't actually tame, but he referred to it that way after seeing it every few days. One morning, it crossed so close on a sand range that Peterson almost stepped on it. The wolf would've been shot as quick as a wink, but he could never get his dogs stopped and the rifle out in time.

One morning, we met on the trail and he invited me to spend the night with him. His camp was only eleven miles away and, since it was Christmas, I took him up on the offer. To prevent the dogs from barking and raising hell all the way, I suggested he go on ahead and that I'd catch up with him later. After giving him a half hour start, I came along behind.

Following his trail was easy, and soon I was traveling on a large muskeg beside a big hill. After going only a short distance along it, I saw a big wolf suddenly come off the hillside immediately in front of the dogs. This was an exceptionally good team and when I yelled, "whoa," they came to a sudden stop. After pulling the head line tight and taking two or three turns around the lazy back, I grabbed the .270 Winchester and waited. Knowing the wolf would stop when it arrived at the top of the other ridge, I got ready. It was a long shot, but I didn't worry because he was right out in the open and could be seen quite clearly. The dogs waited and watched eagerly as the wolf came to a stop, turning broadside to look back. I lined up on him and fired. The wolf made a couple of jumps and keeled over. The dogs were tired of waiting, so we beat it over to him. He was a big gray.

I'd plugged right through the shoulders. I picked him up and placed him on the sleigh. I drove back to Peterson's camp in no hurry. Skinning the wolf could wait until later and the rest of the trip was to be enjoyed.

When I reached the camp, Peterson had already finished his dinner, and wondered where I'd been for so long. I told him there had been a wolf hunt and that wolf of his had bitten the dust. He displayed some surprise but I told him not to worry because we'd enjoy a bottle of liquor on him when we returned to Edmonton.

Peterson was only joking about being hurt by my killing the wolf and said he would've shot it himself, but was never ready at the right time. Still, we had quite a time joking about it, with him trying to act hurt as if I'd done in one of his dog team.

I had an opportunity to observe a tame wolf one time when Hughie placed a couple of traps around a caribou cache, and caught a wolf in each. One he killed by hitting it on the head with his ax; the other showed little sign of serious injury, and so he came up with the wild idea of keeping it and training it as part of his team.

After a real battle of wills, he finally succeeded in breaking the danged thing and drove it as part of his team for the rest of the winter. Since it chewed through leather like it was butter, chain traces had to be used. It was a one-man wolf, with Hughie the only man able to handle it. Dogs can rarely get along with wolves, but since this was a female, they tolerated it. Surprisingly, that thing became an A-one worker, but ate like a horse. If I remember correctly, the wolf was taken back to Snowdrift in the spring with the intention of breeding her with a dog. When she was unable to conceive and looked so miserable tied up all the time, Hughie showed real heart and turned it loose.

In the years we were stretched out across the tundra, predator control was a working success and the wolf population was kept well in check. Once, an announcer from the radio station in Yellowknife interviewed me about wolves, and I let my stand be known telling things like they really are out here. The figures I have are very close to the tally of wolves taken the first four years, and they went something like this: 1,400 the first year, followed by 800, 600, and 400.

As these figures show, the white trappers on predator control were taking care of the wolves, and the caribou numbers were increasing every year, in turn. For example, I saw more caribou in 1953 than in any year that I can remember. That goes back to 1930. Now there is only one white trapper out there, and he is getting too old and thinking about giving it up before long. When we took care of the wolves, we averaged around fifty apiece. Now, the wolf population will go sky high, and the caribou population will drop off accordingly.

CHAPTER XIII

The animal with the distinction of being the toughest and quickest animal walking the North Country was the wolverine. He could make shambles of a trap line and reduce a cache of caribou to a worthless, stinking pile of meat and bones. Wrecking a cabin or a caribou skin tent just for the sake of enjoyment fit his character perfectly. With more guts than anything God ever created, he could cause a man six months of misery with only thirty minutes of mischief.

Even though he was dreaded around a trap line, the wolverine was nevertheless enjoyable to trap because he matched wits with a man. This was a creature who was awfully cock-sure of himself, but he did make some mistakes and, armed with a little knowledge of his weaknesses, a trapper could entice him into stepping on a trap spring.

But if the wolverine was to be held once he was caught, it was absolutely necessary to use a big trap. Occasionally, fortune would smile, and he'd stay put in a fox trap, but I never counted on this happening. In most cases, he would either twist around and pull out of the small trap or chew his foot off. That's how tough he was. It took a big trap like a number four, which grabbed him up high with wide jaws, to effectively get the job done.

Any time I noted any sign of his presence around my line, I went after him with the full intention of putting an end to him as soon as possible. I didn't run out and start setting traps without first taking time to learn his territory. Then I would find a way to hang a bait well above ground level, but far enough below a limb or pole so that he couldn't climb down to it. One of the secrets to catching wolverines was to make them so mad they'd become careless.

After getting the bait situated, I'd use a low fence of wood to herd the wolverine toward the bait. I'd cut some boughs and use them to build a runway on each side of the trap directly below the bait. Once that little glutton spied the bait, he forgot about everything else, and walked down the chute. Then I had him. When the jaws of the trap started together, all hell would break loose. He would fight for his life desperately and wouldn't succumb to become a source of fur until he froze to death or was killed by the trapper. For some reason, I had very little trouble catching wolverines. Maybe I got them because I would go out of my way with my efforts, all too aware of what trouble these guys could cause.

Accident and coincidence played a part in some of the wolverines I caught. A few years ago, I bought a freshly killed moose from a native for thirty dollars. He'd been out hunting, and had come face to face with two moose. He'd bagged them both. Since I was staying in the timber at the time, I had no access to caribou, and so the moose meat was important.

After making our bargain, he told me where the moose lay. I struck out immediately for them, and came across a set of wolverine tracks leading up to the site where he'd gotten the kill. After skinning and quartering the moose, I decided to set a trap before heading back to the cabin. When I returned to the moose kill, I found a wolverine securely in the trap. I set the trap again and, when next I returned to check it, found another wolverine in its jaws. This was a very unusual happening and one of the best deals I ever made. The thirty dollars I'd shelled out for the moose meat was pretty steep, but the addition of the two wolverine pelts made it a very good deal. They were worth a great deal more than thirty dollars, plus I had a good supply of meat for the winter.

The wolverine was one ornery animal, and really got around in the country. One fall near Aylmer Lake out in the Barrens, I killed some caribou which I gutted and spread out, belly down, on the ground. I made the usual cache—with a layer of brush heaped over the meat and a pile of rocks stacked on top of it all—soon after getting to my camp during the caribou run in September. When I needed some meat sometime in November, I found this cache opened. Unknown to me, a wolverine, finding the heap of rocks on top too much for him to move, had tunneled under the protective rocks and was still inside. I started clearing the rock layer. Pretty soon, my shovel broke through into a tunnel which was yellowish with urine. By the smell alone, I immediately knew this was the work of a wolverine, but it never once occurred to me that the critter would still be lurking inside. If I'd even considered this possibility, I would by no means have been digging around in that hole to my chest. I collected two quarters from the cache for the dogs.

The dogs barked and growled all that night with a tone of excitement that indicated the presence of something very close. At times, they would really go at it, then stop for a half hour or so before barking some more. This continued until I got up at daylight. Since there were plenty of wolf tracks around and I had spotted one on the way in, I believed they were cussing wolves.

This was at my second out-camp and my plan that morning had been to run the half-day line and return at night to sleep. Since I was returning, there was no reason to take dog feed along. Instead, I went to the cache for two more quarters, intending to throw them into the tent so they would thaw sufficiently to cut with the ax on my return.

At the cache, I again shoveled snow away to get to the meat. This time, all at once the wolverine poked his head from the hole, startling me half to death. The dogs again perked up and started barking excitedly. I told them to hold their horses, because we were going to get him.

With this discovery, the mystery of their continual barking the night before was revealed. I knew the wolverine had been trying to leave the cache all that time, but hadn't been able to because the dogs were tied so close. He most certainly wasn't afraid of the dogs, but all the noise had made him decide to remain inside. Apparently, they had barked every time the bugger had peeked his head from the hole, and stopping only when he retreated.

Not wanting him to escape, I did the only thing I could think of at the time, and started shoveling snow back into the hole. I kept it up until the hole was completely filled and then eased away to get my rifle. I returned to the cache and, very carefully, started scooping snow away once more. I cleaned it for the third time, and then backed away with the rifle ready. Sure enough, his head appeared above the hole pretty soon. I blew it off.

Stink! Never have I smelled anything so foul. It was the worst odor in the world. He'd been down there for several weeks, maybe since soon after I'd made the cache. There was no reason for him to move out because all the meat and comfort he required were right there. He'd spread urine over most of the meat so that other animals would stay clear of it. That's one of the wolverine's old tricks and it usually works.

A wolverine's short-legged, clumsy appearance may suggest otherwise, but he is faster than greased lightning. So fast in fact, that he will kill a wolf or a caribou and each bloody well knows it. Many consider the wolf to be king of the beasts up here, but I've seen scenes where a pack was feeding on a caribou only to get the hell out of there when a wolverine stopped by.

One will hit a trap line with the destructive force of a twister, never missing a bait or a fur. One time on one of my lines, a bloody wolverine ruined twenty-two white fox furs in one running. He was a complete devil, eating a few, carrying some away and tearing the rest to shreds. I got after this adversary with a vengeance, and put an end to the problem forever.

No matter how many caribou were cached, wolf and fox carcasses were saved for dog food. At one of my out-camps, I was storing some fox carcasses inside a skin tent. A danged wolverine got in and stole the entire works. Knowing he'd be back because the culprit usually returns to the scene of his crime, Delphine and I decided to fix that bugger for good. Before leaving that camp, we hung a fox on the front of the tent in plain sight. Below it, we set two traps with the chains far enough apart so he'd be caught by two feet. Finally, we built a runway that led up to the traps. When we returned to this camp a few days later, the wolverine had become fur. He was caught by two feet and was frozen stiff.

Out in the Barrens, this two-trap method was the only sure way to hold this customer. In the trees to the south, one could use a little different approach because of the amount of brush and poles available. Here, you might anchor a single trap to a long toggle. Unlike a solid stake, the toggle moved with the animal, preventing him from pulling out of the trap and eventually getting tangled up some place. When a wolverine was caught this way, he would drag the toggle around with him, making him easy to follow and limiting the extent of his escape. Any wolverine caught this way was a live wire, hissing and raising hell like a demon.

This loose toggle method couldn't be used on the tundra because there were no trees around to slow a wolverine down. Out there, the trap chain had to be frozen down, and that gave him an opportunity to use his strength against something solid. And this strong bugger could twist a trap chain in two.

I got a show of this animal's strength in the early days while I was still south of Great Slave Lake. It involved a particularly lucky, or particularly smart, wolverine who was running my trap line, tearing up pelts and springing traps as well as stealing baits. I couldn't seem to get him in a trap, but had to do something because the rascal was destroying the entire winter's effort. I placed a tempting bait in a place where the wolverine could find it, but not reach it. Beside it, I rigged a steel cable that would serve as a snare. There wasn't a chance for him to get to the bait and unlike a trap, the snare couldn't be sprung. All the wolverine had to do was get mad enough to become careless, then perhaps he would get entangled in the snare.

Sometime later when I checked the set, the scene bore evidence that the wolverine had been there and had gotten caught in the snare. But thanks to the kind of power only a wolverine can generate, the cable had broken. It was a steel cable, mind you, and during his struggle to break it, he'd torn up everything around the set.

This wasn't the last time I was to meet this tough customer, either. The next year, I caught a wolverine in a typical set and thought nothing much about it until I was skinning the animal. It was then that I found the noose of the steel cable from the snare the year before around his neck, nearly cutting through the skin. After that devil had broken the cable, he had worn the noose around his neck for a year. Yes, he is one tough customer, and any trapper who finds one in his territory soon knows it. Usually, he finds out the hard way.

One year at our place on Beaverhill Lake, we had one of our best runs on white foxes, taking a whopping thirty pelts on a one-day line. There were too many for the sleigh, so we had to leave some at the out-camp, which was a skin tent. Near the end of the line, we came across the tracks of one of those masked bandits. This presented a problem for the fur left behind, because he would smell the white foxes and go after them in a hurry. It would be nothing for him to get into the tent and carry everything away.

We decided to set a trap, but couldn't find any big ones. In fact, the only one around was a number two, but I was pretty sure he wouldn't get out of it once he stepped on the trigger. Close to the tent, not more than fifteen feet away, was a natural place for the trap.

The tent was located on the side of a ridge, but somehow we were fortunate that the thing didn't become snowed under. The place we chose for the trap was a natural, because the wind kept the snow from drifting and collecting there. We placed a couple of frozen white fox carcasses out with only their backs exposed above the snow. I melted ice inside the tent and poured the water over the toggle, to make sure it was a solid set.

The next time we came to that camp, we had our wolverine. Fortunately, he had found the trap before taking time to cut into the tent. If he had gotten in there, the white fox furs left behind would've been ruined and most everything else would've been carried away. As it was, everything remained as it had been, and we picked up a valuable piece of fur.

Along the Arctic coast, the Eskimos got little or no wolverine fur and so liked to trade for it. It was very important to them because, unlike wolf or fox fur, frost won't collect on it. Since the moisture in the breath won't stick to it, wolverine fur was a favorite for lining the inside of parka hoods. I still have an open trading channel shipping both wolverine skins and moose hides to the Eskimos in exchange for seal skins.

In a roundabout way, the wolverine was responsible for one of the most peculiar stories ever published about the North. It originated from the imagination of a writer who tried to piece together the story of an abandoned cabin without learning the facts first. But when I think about it, the whole thing must have looked mysterious to the eyes of an outsider.

At one time, there was a surveyor who was sent to work in the Churchill area. His wife, who was a writer, accompanied him while he worked. To the best of my knowledge, they traveled around the country by helicopter. While flying along the Thelon River, they spotted a cabin and landed the machine for the purpose of investigating it. There are very few buildings at the edge of and beyond the tree line, so this was a natural thing to do.

They came upon the place to look around, and found it deserted. The woman's curious eye was quick to discover an old shotgun with some assorted shells around. Thinking that someone had possibly died there, she looked for evidence of a human being, but found none. The mystery of it all led her to the conclusion that something was badly wrong, and labeled the place a ghost cabin. In fact, I think she labeled the article she wrote, "Ghost Cabin of the Barrens."

Someone sent me the article and I got quite a kick out of it. It was well written, and everything sounded appropriately mysterious, but it failed to include the true explanation for it all. Of course it was very natural for a visitor to make mistakes with no one around to explain things. Few people know how we lived out in the Barrens, or how we had to make do with the limited provisions we had. One of the photographs with the story showed a wheel on the ceiling of the cabin. The writer was puzzled about that wheel and the purpose it served on the ceiling.

I knew the cabin well—who built it, who lived in it—and had spent several nights in it myself. At the time those folks stopped in, the camp would naturally be deserted because it was in the summer while the trapper was out to the South. If he came back, it wouldn't be until September. The gun had been left behind, even though it worked fine, because it wasn't worth enough to carry back and forth.

As for the wheel, it was part of a two-wheeled cart used to pull things by dog team across the tundra in the fall before the snows came. A toboggan was attached to the frame above the wheels and the dogs pulled it around. Since it had a tendency to tip with nearly every bump, it was used for only a short time. The cart was discarded, but one of the wheels was taken to the cabin and suspended from the ceiling to serve as a cache. It was the only thing foolproof to a wolverine. He couldn't get to it, or to the salt, sugar, rice, beans, and tea that were left there in case someone came along needing food. Most trappers tried to leave a little emergency food at all cabins, hoping a wolverine wouldn't find it.

Wolverines are by no means the only pest out this way. Bears can be troublesome at times, although seldom does one have to be destroyed. A surprised sow with cubs can be problematic if a man should find her in the wrong place.

Mr. and Mrs. Art Meinke have a cabin down on the point near my home, and come in for six or eight weeks during the summer. A couple of years ago they had a true bear experience. A big old black took after Art while he was trying to frighten it away. The darned thing came right up on the porch after him, and Art had no choice but to shoot him. Right by his front door, no less.

Bears are fine animals who deserve their place of existence here in the North, and I don't think they should be killed unless it is absolutely necessary. They cause very little trouble beyond occasionally appearing as a pest at camp looking for a hand-out. Mind you, when dogs are around, this happens very seldom.

Black bears aren't restricted to the trees, and they sometimes go out into the Barrens for quite a distance. Occasionally, those out there become bad ones, too. Jack Knox was once in a circumstance where he had to shoot a big old black. He had just killed two caribou and was waiting for a while before going over to get them. When he did, he found one of them was missing. He searched around and discovered a trail over which a bear had dragged the caribou. By the size of the tracks, he could tell it had been a huge bear. Not caring to lose his meat, Knox had decided to stay around close to the other caribou all that night. During the evening, his dogs started barking and woke him up. Looking out, he saw the bear who had returned for the other caribou. The darned thing had no intentions of stopping, so Knox had to kill it. That occurred almost 150 miles beyond the timber.

There are grizzly bears out on the tundra, too, although they are smaller than those found in the mountains. They seem to be very scattered, but are by no means scarce. They don't see very well, but will come to investigate most anything that moves, and have a very keen sense of smell. They are a very strong animal, and can run almost as fast as a caribou, although not as far.

One year, I spotted three musk ox at the end of Artillery Lake with signs indicating that a big grizzly bear had been after them. I followed the tracks for a long ways, but the bear had either smelled me or heard the dogs, and gone across country. He went across a river on ice which was too thin for me to try. He was a big son of a gun, and how he made it over without going through is beyond me.

Although I had to shoot a bear one time, I didn't want to; the circumstances at the time left no choice. Delphine and I took a native boy out with us one year. He was out hunting one day and had succeeded in killing two caribou. For some reason, he failed to make a cache, figuring to return the next day. When he went back, though, a big surprise awaited him. This big grizzly bear was on the scene with the boy pitching poles, cans, and everything else he carried. The boy ran like hell for camp and was scared half to death when he got in. When I went back with him, I found the meat that the bear had covered with vegetation. I also found the bear's cache. Then I looked up, and there he was. He got plenty close enough for me, and not appearing like he was going to stop, I put four .250 slugs into him.

I knew he was hit with every shot, but all he did was turn away and amble off. The boy threw a shot at him, but was too scared to hit anything. The bear disappeared among some rocks and brush. There was no way I was going in there after him, and the boy, similarly, required no urging to stay back with me. I had told the boy not to shoot if the bear showed himself, planning to do all the shooting myself. But he became excited when the bear came out, and fired anyway. After he missed and the bear went over a hill, I got mad.

Later on, I saw him standing upright between two small lakes. Getting as close to him as I safely could, I plugged him again and he dropped immediately. I was afraid to approach him right away, so I waited. When I finally did go over to him, he was stone dead. He was too big for the two of us to turn over for skinning, so we had to skin a quarter, then cut it away. Eventually, he was light enough to handle and we finished the job. That was one of the few times in my life when I was so scared that cold shivers ran down my back. That bear had to weigh over a thousand pounds.

Another thing, if a grizzly had to be killed, the government demanded the hide, skull, and, if it was a male, the trademark. There is a terrific fine if a man gets caught violating this law, because so little is known about the Barren Land grizzly and its habits. Once at one of my caribou skin tents at an out-camp, one of the buggers went in through the door, but came out another way. He was probably confused, because he'd cut a hole through the back like a person would with a pair of scissors. I wasn't there at the time, and it's probably a good thing.

I never knew of any polar bears coming as far inland as my trapping territory, and know very little about them. One trapper insisted that he saw one on Beaverhill Lake in the early days, but I doubt it very much. Perhaps he saw an albino black or grizzly. I know for a fact that polar bear hides are sky-high at the fur auction, with one selling for $1,954 last year. The same fur four years ago was worth no more than $400.

One pest I always had it in for was the gull. Out in the Barrens, this bird was an almost intolerable nuisance. He always found whatever caribou a man didn't get cached because of weather or darkness, beating you to it with a bunch of his buddies, no matter how early you got up to finish the job. With his keen eyesight, he could tell what was going on before any man. The bloody things would find the meat, pick the ribs clean, then crap all over the rest of it.

By golly, a danged gull isn't impossible to eat. Mind you, I never intended to eat one, but was traveling with two natives from Fort Reliance to Yellowknife once, and we ran out of food. On shore one night, I had begun to boil some tea when one of the natives asked to borrow my shotgun. This surprised me, but when the man was asked why, he said he was going to get some meat. I doubted it, but delivered the gun anyway. That guy began shooting gulls, young spotted ones only, stopping when five or six were down. He cleaned them, boiled the meat with salt and butter, and then cooked it. I grabbed down into the pot and came out with something looking like a leg. By golly, it was good, and I wasn't long in going back for more. It beat any duck, goose, or even chicken I ever ate.

When I asked the native why he hadn't killed any old birds, he said they were spared to go back south to eat more white men. He was serious, but I laughed anyway. He had confused the gulls with vultures. Gulls also provided the natives with a few eggs which they ate as readily as those of ducks and geese.

CHAPTER XIV

As time passes and a man grows steadily older, he should, by all standards, become smarter with each passing day as the wealth of experiences upon which he might draw increases. But occasionally, the opposite happens, and a man gets careless and takes too much for granted, and ends up doing the same thing year after year. It's about the time he thinks he has everything figured out that a man typically gets caught up with and a big mess clobbers him right in the face. This has happened to me a few times, and I'm fortunate to be able to look back on these instances which, in some cases, could have spelled my end.

In one of my later years, I again got lost because I wasn't watching my step. It was in the spring, while I was on the way back to Fort Reliance from Beaverhill Lake. Because of a severely frostbitten face she'd suffered the year before, Delphine had stayed behind to try and get straightened out. She was taking treatments, but was never completely cured. From the time we hooked up until the end, this was the only winter we were apart.

There was a point on this trip where, if I'd listened to my lead dog, he would've led me to a warm cabin that would've indicated my exact location. Instead, I was too stubborn to let him have his way and got tangled up. And before arriving at Fort Reliance, I had to kill four pups for food because I got too far south before coming to my senses. Before leaving the cabin on Beaverhill Lake, I'd made sure enough dog food was loaded on the sleigh to last the trip in. This, along with a good bunch of fur and my camping outfit made for quite a load on the sleigh.

Soon after breaking from my camp, the weather turned very dirty. When it became obvious that travel was going to be impossible, I'd pitched the tent and holed up against the blizzard that was raging outside.

In all, I had to wait two or three days for the thing to blow itself out. Having enough fuel for cooking and warmth was pretty important, and I knew where to find it. As such, I had plenty of wood during my wait for the blizzard to pass. But when a man has to stop on the trail and the stay drags into days, the dog feed starts to dwindle in a hurry. So it was in this case, and I was getting into trouble.

This trip was only 140 air miles, but I had to make a lot more than that with an erratic course that took advantage of the scattered wood patches I needed for camping every night.

After the storm had run its course, I put everything back in order for traveling, and again hit the trail. Since the meat was running out, I was in a big hurry. Sport, a big black Husky who was my leader at that time, knew the country well. Even though a stubborn cuss, he was a good son of a gun who knew well the importance of following a trail. We soon came to a lake that I wasn't able to recognize and, immediately after heading out on it, Sport began pulling away from the direction I wanted to go. Even though ordered otherwise, he paid no attention, and continued to bear to the right. I had to use the whip to turn him. I didn't hit him, but ran alongside, grabbed the trace, cracking the whip two or three times, and yelling at him.

When he looked back at me, he was wild in the eyes and I could tell he was mad, but he turned anyway. Since he had been here before, he knew we were close to Whitefish Lake—only four miles away, to be exact. I had pulled him off the trail because I thought he was wrong. But I was the mistaken one, and my stubbornness sent me to a bloody sand range I'd never seen before in my life.

This trip took place in the first week in March, a month sooner than usual. I'd had to leave my camp and the trap line because the dog feed had nearly been gone. The caribou were spotty during the fall, making it impossible to kill enough to last the whole season. I had worked hard getting a few fish, but, still, there wasn't enough.

Besides the dog team, I had four pups Delphine was raising along with us, and they ate like horses, putting an extra burden on the already limited food supply. Besides the few caribou and fish, the dogs ate everything I could catch: foxes, wolves, and anything else. Of course, with the caribou elsewhere, wolves were also scarce. Most of the ninety furs I had on the sleigh were foxes.

I had waited at the cabin as long as possible in the hope that the caribou would return. When I was down to my last cached caribou, I had to strike out. Neither the dogs nor I were in bad shape, but the trapping season was cut short. I thought of all sorts of things when hitting that strange sand range. Chasing Sport off the trail had been a bad choice and that night I was pretty well disgusted with myself. I pitched the tent, slept in strange country, and the dog food was all but gone.

The next morning, I hitched the dogs up and we again took off. In a short time, we came to a creek. For a while, I couldn't figure the thing out because the water under the ice was flowing north and I was going west. I carried a map which was four miles to the inch, and it showed the stream running north. The whole thing didn't look right and believing the map was almost impossible. I put the tent up right near the creek bank while I thought about the whole confusing mess. At last, I put on snowshoes—knowing they'd provide a quicker means of travel than the dogs could—and went out to see where this creek flowed. I walked for probably three miles over some very difficult footing. The snow was deep along the creek bank which was extremely steep in many places. Overflow ice on the creek surface added a further hazard to the tough walking.

Although the hike was a strenuous one, I kept plugging away until I arrived at a big lake. West of it, a great big range of hills rose up like a mountain. Along with the north-flowing creek, this provided a pretty accurate landmark. Once again, I took the map out and studied it closely. Everything seemed to fit together. True to form, a lake was shown right where I was—a big lake at the head of the Snowdrift River. This had to be the one, I thought, because everything checked out perfectly.

So I headed back toward the tent and, when I was almost there, I discovered the stump of a tree. It was right on the edge of the creek bank, and I had failed to notice it before. Since the tree had been cut with a saw, I figured the work to have been done by a white man building a cabin in the area. Now that I had an idea where I was, I remembered a man by the name of Red Pearson who had located here. I searched the area carefully, and found the cabin. It was built so low to the ground that it looked like a dog house. Sometimes, these crude domiciles were constructed this way because trees were scarce and scattered. At other times, the winter season approached too fast and cabins had to be built hastily. I never found out what had happened to this one, but it sure had a low bridge. I later learned that an Indian family had occupied it for a time while trapping. When I was there, the thing hadn't been used for years.

I broke camp and made tracks along my trail to the big lake. I saw a world of caribou trails along the way, but they had already passed through headed north to the Barrens. Wolves howled every night, but were far away pursuing the caribou, and I never caught a glimpse of one. A wolf carcass would've provided food for the dogs, but I couldn't shoot what couldn't be seen. By this time the meat supply was gone, and the dogs were feeding on my own eating meat. I was now down to salt and tea and, having only seen one fresh caribou track, I knew the prospects for finding meat were dim.

Out of desperation at the tent that night, I killed Diamond, one of the four pups. I shot him with the shotgun I'd carried to hunt ptarmigan. There had been thousands of these birds but, like the caribou and wolves, they'd already passed through to the Barrens beyond. It became pretty discouraging to see all sorts of game tracks when you were hungry without being able to shoot a single animal.

I skinned the pup, and, all except the ribs which I kept for myself, fed it to the dogs. Since they were so hungry, they wolfed the meat down at once. But it was too warm for their systems and they threw it up immediately. Accustomed to eating only frozen meat all winter long, the sudden change had caused a reaction in their stomachs. Since there was no other food, I picked the stuff up and placed it on the back of the sleigh the next morning; after it froze, this would be fed again that night.

The dogs were getting weak and I had to push them. They eventually found out that I meant business and went ahead, but I could tell they were having a hard time of it by the amount of barking and howling they did. This was not like them, because they were a good team. I sure felt sorry for those guys, but had to continue to coax them along.

Much of the time on this trip, the snow was very deep—so deep that the leader wasn't able to jump up on top to wallow it down—and I had to go ahead on snowshoes to break trails. Any man who has broken trail in this kind of deep snow knows how hard the work is. At times, you feel like saying to hell with it and giving up.

After I finally came to the big lake beyond the cabin I had scouted earlier, it became a little easier for the dogs and they began to pull better. I kept going, having decided to follow the map as closely as possible. At the end of the lake, I came to a river which was flowing west. This filled me with relief. I knew things were right and that this had to be the Snowdrift River—but an occasional doubt still found its way around in my mind. This happens when a man has been lost for several days, during which time many things look familiar but eventually prove otherwise. If I kept on my present course, the final proof of my location would reveal itself shortly.

The first year I ever went into this part of the country, three of us had gone over Pikes Portage, making three lakes east to Little Whitefish Lake. From there, we'd branched off to hit Wolf Creek, which we'd followed to Quartz Lake and eventually to the Snowdrift River. We'd gone down the river in our canoes for a couple of miles, then chawed up the Eileen River to Tent Lake. It was only about ten miles to the Barrens from there. Hughie was along and knew the bloody country like the back of his hand. I learned much of what I do from him.

While traveling along on this journey, I saw more and more evidence of caribou, but the trails were old. I kept looking for stragglers, but it seemed as though they had all gone out to the Barrens at about the same time. One hazy and foggy day, I ran upon a big bull, but before I could clear the rifle, and in spite of the fact that the dogs had stopped for me to shoot, he was gone, never to be seen again. We gave chase, but the going was too rough with the big load and we couldn't get close.

I kept following this river which flowed west from the big lake, wondering when I'd recognize something I'd seen before. The going got tougher and the dogs got weaker by the mile. I was running out of time rapidly, but those dogs wouldn't give up.

Finally, I came to a creek and left the river to follow it. If things worked out right, in a very few miles I'd be in familiar country and the map could be put away for keeps. It couldn't be very far, I told myself. Not more than three or four turns according to the map, and there should be a falls at the end of a lead-in river. Then I'd have it licked. Still, I wasn't completely sure of myself until I heard the falls to the left. It was a small falls that probably had been open all winter. They marked the Eileen River, which was the bugger I'd been looking for.

Now that I knew where I was, it should have been clear sailing all the way to Fort Reliance. But the snow was really piled up—about the worst I'd ever seen. In many places, I could make only two or three miles a day. Most of the time, I was breaking trail ahead of the dogs. I'd get up at the streak of dawn, put on my snowshoes, and strike out, leaving the dogs behind and the tent still standing. I went as far as possible breaking trail, then returned to the camp. On the morning following my discovery of the falls, I'd hooked up and driven over the opened trail I'd just made. When I arriving at the end, I made camp once more. This kept up until I got out of the deep snow.

I had to kill a dog—one of the pups that was running loose—every other day to feed both myself and the team. I was always hoping to get a shot at something, but I always failed, and so had to kill those pups to get back in. I was eating those dogs, too. In fact, I ate most of the ribs from all four of them, putting them into a pot of boiling water with salt and rice. And, cripes, I could have eaten the whole team before I got in. I hated to kill those pups, but a man has to live. When a man gets into a jam, or gets stuck with his back to the wall, he has to do something and danged quick. This time, I had no other choice but to hurt and destroy the things I loved—something I wouldn't have done at any other time in my life. The pups were around five months old and beauties to behold. They were not part of the team, and ran loose behind us. Their mother was White Kaviook. She died during my last trapping trip while I was picking up my traps. It burned me up to have killed those pups, and I've had to live with their memory ever since.

This adventure took place just three or four years before I had to quit trapping, so I was no spring chicken. But I was still pretty tough and bull-headed enough to keep going. I had been around long enough to know I would make it if I kept my head and wits.

In all, that trip took me two weeks to complete and I was a very lucky man to live to complete it. This is danged rough country, even for those of us who have lived and worked in it. The very nature of the country has caused some very cruel things to happen, especially to the foolhardy who've taken it for granted. And anyone who thinks the country isn't still rough and wild has another think coming. It can be mean and ruthless to anyone who accepts it for less than face value. A few years ago, some fellows from the States found this out the hard way—and darned near lost their lives as a result.

These four husky chaps had taken off on what they expected to be a pleasurable wilderness trip, and had failed to file an itinerary with the RCMP. That in itself was a foolish thing to do because, in the event that they got lost, the Mounties would never know where to look. Anyway, these fellows were in an eighteen-foot canoe, probably greatly overloaded, and were going down the Taltson River. Before getting very far, they hit some rough water and the canoe swamped. Three of the group swam to an island and waded ashore while the fourth clung to the canoe and was swept away by the current. Some eight miles below the swamping spot, the canoe drifted close to another island and the guy left it and swam ashore.

Several days later, my nephew, who's a pilot, was flying in the area of the Taltson River and spotted the canoe below with no one in it. Most pilots were trained to check all canoes out in the remote wilds, and a quick look told a tale of trouble below. He circled the area but saw no sign of people. He flew on home that night, but couldn't sleep: the thought of that abandoned canoe wouldn't leave his mind.

The following morning with plenty of gasoline and food, he took off in the plane for another look. He located the empty canoe again, and circled the area for a long time. But he saw nothing. Just before giving up, he decided to have one last look upstream. About twenty miles above the canoe, he came to this island. Glancing down to a bald point, he could see two men standing near the water, waving their shirts.

He was able to put the plane down and get close to the men. It was the same three canoeists—now starving and driven nearly crazy by the hordes of insects. Their faces and bodies were swollen and discolored from bites and exposure. They hadn't had a fire all this time, so he built one for them and fed them lightly. They mentioned their missing buddy to my nephew, telling him he'd been swept away and drowned. My nephew wasn't satisfied with their explanation, and refused to accept their suggestion that this other man was dead. Leaving them on the island, he took to the air again to do some more searching. Sure enough, he found the man, who was still alive. When he was picked up, he was out of his mind.

For his part, he wouldn't believe the others were still alive until the plane picked them up. The four of them were lucky to get out of that spot after being lost and marooned for ten days with absolutely nothing for comfort. When they pulled out of the territories, they told my nephew that he would hear from them again.

Not thinking much about it, he was surprised to later receive some papers to sign from a bank in the States. Those four men had set up a fund of some quarter of a million dollars in trusts and bonds payable to my nephew at age twenty-five. He had saved those four guys and they full well showed their appreciation. He'd found them just in time, because they were planning to commit suicide by drowning themselves. Yes sir, any doubters about the rigors of this remote country can learn something from this episode.

But despite the great extremes in weather and temperature, along with the millions of acres that are unfit for human habitation, this is a great country unequalled by any on this earth. The feeling I've acquired from mastering it for all these years has given me a lasting satisfaction and a full meaning to my life. I'd be bulling to say it hasn't been trying at times. But if a man loves the out of doors and the solitude the far North offers, he can endure all the hardships. I made it because I was bull-headed, and never had to convince myself that this was the life for me. I knew it was the life I wanted, and so accepted things as they happened.

I felt a sense of pride and accomplishment after coming back from every Barrens winter even when there had been a rough spot or two. Even if I wasn't loaded with fur, I counted each trip a success to my inner self. I used to look back at the North from Great Slave Lake and shake my fist, saying, "You didn't hold me down, you bugger. I didn't give in to you."

Oh, there were times when I doubted myself, and two or three times I swore that if I ever got out of a particularly tight spot, I'd never go back.

But even as I was making these threats, I doubted them. I knew I was too stubborn to quit.

Instead of going out to the South to find what I had, all I had to do was sit tight and listen to others. It was often the words and wonder I saw in the eyes of an outsider that reminded me of what I truly had and took for granted. For example, there used to be a man from the States who flew out here to fish. We became close friends, and used to sit and talk for hours. One time, just before he was to board the plane to fly out, he said he envied me and my way of living. He was a very wealthy man and when I asked him why, he said it was because I lived here and was free with trout right out in front of my cabin, practically at my doorstep. He said I enjoyed the cleanest water and purest air in the world and, even though he had plenty of money, he couldn't afford what I had. Also, he said he had to go back to work the following week.

Those words really cut into me deeply and I never forgot them. That man knew and understood the magic of the existence I'd chosen for myself, but was surrounded by a way of life he couldn't escape. After some serious thoughts, I knew I'd never trade my life for that of any man.

Others have said they feel sorry for me because I've lived way out here away from the world. But, hell, this is still the world—and a pretty special part of it. Besides, I haven't had it so bad. The mail plane that comes in once a month brings in all the news I care to know. And, most nights, I can pick up a station or two on the radio, and that fills in the rest. Another thing, no one is forcing me to live out here. This is simply my home, the one place on earth I most want to live. I've tried other places, but my roots are planted firmly at Fort Reliance. I'm a man who has to be and do the things his mind and heart tell him to.

Seldom does my lifestyle demand much beyond my cabin and grub, but if there's something I want, I get it. I don't worry about sickness, because I'm never exposed to anything. In fact, the only time I ever had a cold was when going to the South. The North is a healthy place and that's probably why I've lived as long as I have.

Indeed, a good many trappers live to a ripe old age, and a few of us old-timers are still hanging on. Living longer than most people can be credited to the way of life we trappers led. We've spent long hours living in the outdoors with plenty of good, healthy exercise, few worries, and little pressure. Unlike the fellow who's worked at a desk and hopped into a car to cross the street, we've operated our bodies in the way the good Lord intended.

No, I was never sick on the Barrens, a place most people wouldn't spend a winter if given the choice. I never took any real medicine out with me, and wouldn't have known what to do with it if I had. There was always a little iodine for cuts and some salts and a hot water bottle for an enema, in case I got indigestion. Neither did I ever hear of a doctor or a plane going in for any trapper because of sickness. We were strong and active and had enough stamina to stay healthy. Even though there were plenty of chances, the devil never got me.

One year at one of my Barrens camps, I suffered some tooth problems. This happened in the forties, before I temporarily lost my trapping license. Strangely enough, they had never bothered me while I was out on the trail, with the cold apparently numbing them. But the story changed in camp where it was warm. There, the darned things drove me out of my mind and I thought I'd go bugs.

I tried everything imaginable to stop the pain, but there was just no answer. Everything seemed to make them that much worse. Again my bull-headedness took over, and I refused to come out of the Barrens. There was a lot of fur that year with a good fox run on, and I wasn't about to leave a good thing. But on the way out that spring, I had never been so glad to see the lob stick marking the portage at the end of Artillery Lake. Lob sticks were made by men who climbed trees, cutting off limbs all the way to the top and leaving only the green branches at the top intact. They could be seen for a long ways, and guided a man to a portage.

Natives were superstitious about lob sticks, which one man prepared and dedicated to another. They believed that when a tree died or fell, the one in whose honor it was made would also die. One trapper I knew made a lob stick and told an Indian it was for him. The Indian got madder than hell when the trapper said he was going out to chop it down.

Anyway when I arrived at Fort Reliance—and I made a catch that year, too—I wasn't long before striking out for Edmonton. Not very bloody long at all. I was going to find a dentist immediately upon arriving. I did, and he yanked the offending teeth out before I even got the chair warm. Pretty soon, he stopped and told me I had four good ones that he'd leave in. I told him to hell with that noise because they hurt as badly as the ones he'd already pulled. I told him to send them along with the rest.

By golly, he shook his head, but took them out anyway. And after he'd done it, he told me I had been right about them because they were decayed. Those teeth had bothered me for so long that I felt better the minute they were gone. That bloody toothache business was no good and this put an end to that problem.

That's the way it's always been, with me taking things as they came up, and I still function that way. Why worry? Even while I was trapping and having a bad year, it was far better than working for wages for the other man. I felt I was farther ahead, feeling that as long as a man had grub and a place to sleep, there couldn't be too much wrong.

This was more or less characteristic of all trappers, an easy-come/easy-go lot. All of us were confident that we could go back to the trap lines the following year and get more if the funds ran out. We didn't worry or fret about much. Another thing, too: I sure never went job hunting in the summer. That season was for taking it easy for two or three months.

The cold was never a source of worry for me, either. Out here, once winter arrives, that was it for a good long time. Maybe there would be a nice day or two, but no great deal of thawing would take place until April or May. It was cold and hard, but we knew that before coming out here. Anyone who worried about the cold was in for trouble, and if he got too cold, it was his own danged fault. Every day had to be faced as a special one and taken for what it offered because a man couldn't change it. Anyone thinking he could, had better stay home by the fire.

CHAPTER XV

If a single trait of character stands out more than any other in me, it would have to be the strong feeling of individualism I've possessed since childhood. Consistent with this feeling, I have always respected the other man's choice of work and way of life, assuming he was both happy with, and good at it. I haven't met many people because of the place I live and work, so it's been easy for me to trust others from the beginning. The deeds of a man should determine his worth; engaging in pre-judgment too many times proves it false.

Respecting and complying with the wishes of the RCMP officers who were stationed at Fort Reliance from 1929 to the early fifties was easy, because they were outstanding men with a feeling for right and wrong. Those fellows were stationed out here for one-to-four-year stints before switching off with someone in a different place. Most of them believed in their work, and did it in such a way that no one objected or resented their actions. A great many of these men were to become very good friends.

But there was this one crackpot who, no matter how hard I tried, I couldn't make friends with. His name isn't important, but he came out here with a chip on his shoulder. He let it be known from the start that he thought we were all a bunch of outlaws. In a strict sense, he was somewhat right, but there were a lot of ways of looking at things. Sometimes, we had to make our own rules out of necessity because, when hunger took over, survival depended upon shooting meat. Shooting caribou between April 1 and September 1 was illegal, and justifiably so. But if a man got caught coming in from the Barrens after April 1 and ran out of meat, what else could he do? We had some arguments about this and in the midst of one of them, he told me to back off because he had no friends and didn't need any.

Well, he proved well enough to me that he couldn't be trusted. His big concern was the Thelon Game Sanctuary on which he figured some of us were trapping illegally. He let it be known that someone was going to be caught and would pay the price.

The sanctuary was established as a place of protection for the musk ox. The biggest problem resulted from the fact that there was no natural boundary to the south—no rivers, lakes or hills. All of us knew about the place and stayed far enough away to stay out of trouble. In fact, I'd been in this same area before and RCMP patrols had stopped in to see me. None had ever mentioned anything about trespassing.

This Mountie, however, kept searching around and, finally through the moccasin telegraph, found out about a white man who was trapping without a license. He was arrested, but was promised to be forgiven if he would tell all he knew about the white trappers on the Barrens: who they were, where they trapped, and anything else he could come up with. This guy told the Mountie anything he wanted to hear so he could get off easy. This trapper had a rough time of it, and all of us had helped him out in one way or another, so his lies came as a surprise when we heard the son of a gun had told the Mountie we were all on the refuge. This really started the ball to rolling.

A special patrol was sent out toward the refuge with a native interpreter leading the way for the Mountie who was afraid of the Barrens anyway. As luck would have it, they came across Phil's and my trail, and followed it to our camp. At the time, we were out on our lines. With our camp vacant and the patrol unable to wait around for us to show up, they headed back to Fort Reliance, taking along some things for evidence: some dog harnesses, a spare rifle, some ammunition, a dog whip, and God knows what else. They left a note telling us to report to Fort Reliance in a week, because we were trapping on the Thelon Game Sanctuary.

When we got back to camp, we found the note but ignored it. The order was the most unreasonable thing I'd ever heard of. There was just no way we could get ready to go on that short a notice, especially since this was the last of March and we had traps yet to pick up and fur yet to fix. We closed our camp the right way, then made the trip to Fort Reliance.

After we got in, they decided to try us in Fort Resolution, which was 300 miles away. We agreed to this, in spite of the fact that our only means of travel was a dog team. We were innocent, and determined that they wouldn't get to first base. We stopped in Snowdrift, where I caught a bad cold that nearly led to pneumonia. I stayed back in Snowdrift, and Phil went on to Fort Resolution. I had intended to catch up later, but the sickness hung on, and the season for dog traveling came to an end.

Phil was tried by the Justice of the Peace. After a couple of days of wrangling and trying to figure out this and that, the J. P. decided he couldn't convict him. The fact that there was no water course or marked boundary defining the Thelon Game Sanctuary along the south edge made the testimony too much a conjecture and not enough fact. The case was thrown out of court.

When it came time for me to go to court, they decided not to take my case before the same judge. They wanted the Inspector to try the case, which was all right with me. I went into Fort Resolution with the police detachment in a police boat. After waiting around for three or four days without anything happening, I told them I wasn't going to stay there all summer, and unless the Inspector came pretty soon, I was going back to Fort Reliance. My case was postponed and the trial was rescheduled for the following summer.

In the fall, I went back to the Barrens to trap the same line I'd trapped the year before. The Mountie asked me where I was going to trap, and I told him the same place he'd found me before because I was legal and he would just have to prove otherwise. Mind you, the only things defining the imaginary line were two stacks of rocks called cairns, one on Crystal Island in Artillery Lake and the other on the Thelon River, some 150 miles away. It was possible to be a few miles inside the Sanctuary and not know it, unless the Hanbury River or some other water route were seen. There are literally hundreds of lakes in that country, and with snow on the ground it's hard to tell where you are.

The following spring, a patrol came to my camp to tell me to report to Fort Reliance because the Inspector was waiting for me. I responded that I'd be in within three or four days, and then picked up the traps and broke camp.

The Inspector, Birch, was a very nice fellow who had been on the job for several years and was ready to retire and move to Vancouver. He was in charge as I was tried at Fort Reliance in what we trappers called a "kangaroo" court. Sitting in on the thing were Jack Knox and one or two other trappers, plus the native interpreter.

In a lot of ways, the Inspector took my part throughout most of the hearing. He called the interpreter up and put a map in front of him with hundreds of lakes on it. He asked the native if he could find any one he wanted to if there was snow on the ground in the winter. The native said he thought he could. The Inspector told him firmly, "You could not. How could you tell if it was this lake or that lake? There are hundreds of them with some being no more than potholes of only two or three acres."

That testimony was tossed out, but the crackpot came forward with some pictures he'd developed from film taken from us for evidence. One of them was of a cabin. He asked me where it was located. Rather sharply, I told him there were hundreds of cabins in the Northwest Territories built of logs and this could be any one of them. I told him if he could prove where this one was, I'd answer any question he could ask.

He turned to the Inspector and said I was committing contempt of court. The Inspector said I was probably protecting someone else and didn't want to involve them. Since I was the one on trial, my answer was acceptable. You know, what he said was pretty well true, but I wasn't going to say anything.

He took my part again, telling me right there that he regretted trying the case. Birch understood how things were with the country to the south of the Sanctuary, and how it was to travel without a large lake or water route to guide you. Under the circumstances, he said he would go easy.

They had taken $200 worth of fur for evidence, and they kept that. Additionally, I was fined $75. The Inspector said the fur was confiscated because of the amount of trouble they'd had to go to with the patrols. He figured the money should go to the Crown, and I didn't object. I asked him if there was any danger of losing my trapping license, and was assured there wasn't, because that fact had nothing to do with the case. He said

I'd been convicted on a technicality, and that my license was intact. I was satisfied and thought nothing more about the case, not even considering an appeal.

In the spring, I went south to Edmonton where, before long, I received a letter from one O. S. Finney of the Deputy Commissioner's office stating that my trapping license had been revoked indefinitely. Anyone can imagine about how I felt, because trapping was my livelihood, year in and year out. The letter stated there was no way to appeal the matter.

It took me three years of fighting and maneuvering to straighten things out well enough to get my license renewed. I wrote the Deputy Commissioner and even ran into him in Yellowknife. He told me he would review the case, but nothing came of it.

Next, I contacted the Inspector at Fort Smith, giving him all the details leading up to the trial. He assured me that he would do all he could to re-open the case. I wrote to Gibson and relayed this to him, but he wasn't overly receptive. He wired back to say that the police didn't like their work undone. But I got a letter from the Inspector at Fort Smith stating that he had recommended that my license be reinstated.

That summer, I went to see the Justice of the Peace who had tried Phil on the same case and thrown it out of court because of circumstantial evidence. I asked him if he might try to do something for me. He was sympathetic, and wrote to his brother who had worked for the federal government, but was now retired in Victoria.

Not long afterwards—in fact only two days before Christmas—I received a wire that delivered all the cheer I needed. My license to trap had been reinstated. By golly, it didn't take long to get my dogs and equipment together for a trip to the Barrens in January to try my luck again. I was once more a free man, but I had also learned some lessons. A funny thing, the government the following year shoved the boundary of the Thelon Game Sanctuary back north fifty miles to a point about ten miles from the Hanbury River. It was probably the prospectors who were responsible for this, but it became open country for trapping, as well.

After getting my license back, I had some fifteen years of trapping the Barren Land remaining. I experienced no further trouble with the law. In fact, after hearing about my case, the Mounties made several additional patrols out toward the Sanctuary. They later told me that they didn't see how my case ever came to trial, because even they couldn't tell when they were on or off the Sanctuary themselves. At that time, there was no aerial map. All they had was a ground map with a scale of thirty-two miles to the inch. Hell, the devil himself couldn't tell where he was with that thing.

This little episode in my life was one over which I had no control once it began to unfold. The feeling of relief I experienced after getting the renewal far outweighed any of the animosity I held. Yet, there were times when I did some danged fool things over which I did have a choice. Perhaps some of these mistakes were necessary, because how else would a man learn?

Even though I knew darned well it wouldn't work out before I plunged head-first into it, I decided to go south to the city to find out how the other end lived. From the experience of my early life and my experience trying to make my marriage work, I knew I wasn't a city man. Sure enough, it wasn't long before I was sorry.

But at the start of this misadventure, I thought I had found another way to strike it rich, and had made a down payment on a fur store in Edmonton. That was in 1943, over twenty years after I'd come into the North Country to trap. I was keen to learn all I could about the fur business, and so had decided I needed to work at the other end. But it didn't take three months for me to learn that I was getting nowhere in a hurry.

Another fellow who'd promised to put up some money to go into the business with me, was supposed to be my partner, but he was smarter than I, and had backed out. That left me holding the bag. Right away, I learned the fur business inside a store was nothing like the trapping business. My freedom was gone, and it took too much money to keep the operation going. I needed to see the sun and stars from horizon to horizon any time I wanted to look. That wasn't possible in a store in Edmonton—as I found out the hard way.

Still, there is a little value in everything a man does, even if he hates the task at the time. It was true in this case, too, because, for one thing, I learned what happens to furs like mine when they finally get to the furrier and eventually to the customer who wears them. Once, a lady brought a piece of fur to the store to have it cleaned. Hell, the furrier cleaned the thing the same way we did in the North. When she left it to be glazed, nothing fancy was done to it at all, with not a single chemical employed in the process. In fact, the only extra thing added to the fur during the whole procedure was water. The fur was thoroughly washed to loosen the particles of dirt, after which it was fastened to a wire and let hang to dry. Then, they simply beat the hell out of it with a stick. With all the dirt and dust removed, the fur looked as good as the day it was new. The customer was then charged five bucks. Nothing to it.

Watching closely, I learned how fur coats were made, and was amazed at the number of pelts it took just to make one coat. For example, one muskrat coat took sixty skins to complete. Seeing that coat take shape told me why the final cost was so high.

To make a coat, the flesh side of the pelt was carefully prepared with every single piece of fat and flesh removed. To treat the skins, they rubbed them thoroughly with sheep tallow. Afterwards, each pelt was ironed by hand—a hot one, too—without any harm to hide or hair. Then it was blocked and blended. Since only the prime backs were used in the coat, the furs were cut jagged with the pieces fitted together. All pelts selected for the coat were cut on the flair.

The pelts were sewn together and the block of fur was put on a table where it was stretched tightly in all directions by nailing it with copper nails all the way around. With a piece of chalk, the furrier then marked it up from a pattern before sending it out for tanning.

When the tanner had finished, the fur was shipped back and the pattern was cut out. Using a mannequin, sewers sewed it together. A good furrier could do all this and have the coat ready in about two days. He was quite fussy about his work. If a sleeve was even a quarter inch off, the stitching was torn out and the thing was redone to perfection.

A big surprise came when I learned that furs could be kept for several years when properly fleshed and dried away from excessive heat and stored in a cool, dry place. They had to use moth balls and the furs had to be kept clean. This made the rumor that I had always heard about big fur companies holding onto furs for years likely. I could see how these cheap furs could be put on a high market later, and that a fortune could be made.

I also saw the importance of a good fleshing job and realized the value of a good torpedo when it was properly used. While the skinning was done out in the bush, the fat was frequently washed from the skinner's hands with soap and water. Rubbing too much grease from the carcass on the fur side of the pelt would ruin it. Now I knew why the furrier had to have a nearly perfect pelt at his end.

Out in the wilds, foxes became rubbed in the spring on their hips from sitting so much. I used to spend hours on end preparing these late foxes. I'd cut a "V" from the pelt to eliminate the worn-down area, then sew it together to avoid being docked by the buyer. There was really no problem, because the buyer was most interested in width, as was the furrier. But if the furs were sent in without being fixed, they would bring less at the auction. It was a trick of the trade that made me some money over the years. I felt better knowing that the furrier wasn't cheated by the doctored furs.

Even though the fur business at the final end held nothing for me, my exposure to it taught me to respect the fine work of those people. Also, it let me know that my part, the fur-taking business, was the fun part. Theirs was close and tedious work—and hard, too. Heck, all I had to do was set a few traps and enjoy myself.

Soon, I'd had my fill of the city. I forfeited my $2,500 and headed back to my true home, the North. This time had been like all the rest. My mind had stayed in the North and I couldn't function anywhere else, no matter how hard I tried. This marked the last time I either tried to live in the South or invested money in some wild scheme. Never was further proof needed that I was a man of the North, and the crowded population of the South had made me miserable for the last time. When I went in again in later years, I knew it was only for just a short time. Knowing that there was no chance of my staying was what made it tolerable.

It was pleasing to me to look over the maps of the broad country of the Northwest Territories and see so many lakes named after some of the trappers I'd known. We D'Aoust brothers have our own D'Aoust Lake, but I had never been there until a strange event came about.

One time, a few years ago, I was asked to accompany a flight carrying a caribou survey team along the edge of the tree line to the east of Fort

Reliance. During the course of the trip, we ran head-on into a driving snow storm, and the pilot, unsure of precisely where he was, got turned around. He found a lake below through the blizzard, and landed the plane there to wait out the storm. The landing was an emergency only to the extent that he needed some gas-saving time to get his bearings.

There were plenty of emergency rations on board the plane, but I climbed out immediately to scout around for meat if the need arose. There were rabbit tracks all over the place, and I went back inside to tell everyone we needn't worry about starving to death. The warden in charge was pretty well shook up, but the pilot and I weren't the least bit bothered by the situation, realizing full well that the storm would eventually disappear and that we'd be able to find ourselves from the air.

In due time, our patience proved correct, and the weather cleared up enough to take off. Flying in a circle, the pilot located a river that he recognized right away. He checked his maps and, having made the connection, had a good chuckle. He had gotten seventy miles off course in the storm, and we'd landed on D'Aoust Lake. He laughingly gave me heck because I hadn't been able to recognize my own lake. That was the only time I ever saw that lake, and it had been by accident. Incidentally, we flew north after the mishap, hit McLeod Bay, and eventually returned safely to Fort Reliance.

The North has become famous for many things over the years, but the thing that brings in the most tourists is the fishing, especially for the big lake trout. For a good many years, these fish had gone undisturbed except for natives and trappers who needed their food source for existence. Seldom did either faction fish for sport.

Like everyone else, I have a big fish story. Delphine and I had hooked up the dogs one morning just before the trapping season was to begin, and ran the three-and-a-half miles to Beaverhill Lake. We were hungry for some fresh fish from our net. Well, she cut one of the ice holes open while I cut the other. When we put the running line on and began to pull it up, we could tell immediately that it was heavy with fish. This was late, and we knew it had to be trout. But part of the net came up and it was in a big knot. This was unusual, and I told Delphine there must be a big fellow in it. Before long, we'd pulled it all the way out, and saw that it contained a real whopper, along with a small one. That was the only time I ever got excited about a fish. After taking him to the main camp, Delphine took a picture of him to prove our point to others.

The following summer, a pilot flew out here from one of the fish camps and pulled in to my place to stay overnight before carrying on to Sioux Narrows, Ontario, the following day. He had four lawyers from North Dakota with him. Between frying caribou steaks and preparing the rest of supper that evening, there was part of a bottle of rum.

One of the men spotted the picture of the big trout hanging on the wall and asked how much the fish weighed. I didn't know because there was no scale and no reason for there to be one at a trapping camp in the Barrens. I told him we had fed our seven dogs the trout before leaving Beaverhill Lake for Fort Reliance in April. That night, the lawyers kept estimating the weight of that big trout, and before they got through with him, they had him up to ninety pounds. There's no way to tell now, but compared to the big ones I've seen, this old boy had to be close to eighty pounds. When I held him up, his head was above my waist with the tail dragging the ground.

I have caught hundreds of trout from Great Slave Lake and other lakes out in the Barrens—not with a rod, but in nets. I've never been a rod fisherman, so when I go after them, I use a net. That way I'm sure of getting them. Many of them were large, with several in the twenty-five to thirty-five-pound class. But I've never seen anything approaching that fish's size in all the years I've been in the North.

Another time, a Mr. Orth from Milwaukee was out here on a fishing trip and, when he caught sight of the picture of the big fish, he wanted to know where it had been caught. I told him Beaverhill Lake and that I could show him the exact spot, even though it was 140 miles from Fort Reliance. Jim Magrum was there also, and together we flew out to Beaverhill with Mr. Orth, who wanted to have a go at another monster.

Although there wasn't a ripple on the water, I told him there would probably be no big ones caught because at that period of the summer they were in very deep water. The place I'd caught the big one was in only fifteen or twenty feet of water. While I fixed a shore lunch, he fished for a couple of hours, but only caught around thirty trout with nothing larger than eight pounds. He was a great fisherman, and showed no concern about the lack of size. He just got a kick out of fishing. And there were some caribou in the country then, and we could see some on nearly every high spot. The man was very pleased with the trip to the edge of the Barrens.

I've never objected to sport fishermen coming into the country because they don't take more than they use. Most were looking for a trophy, and followed rules of sportsmanship. Anyway, it doesn't hurt the fish population to harvest a few. But there are other things that men of intelligence do which are Godless and that does bother me. I'm talking especially about the things done outside the North that affect this country. Those things make me sore and I can't understand them.

Like when some outfit put up a big power dam down on the Peace River. I'm not even sure where it was located, though someone once said it was in British Columbia. In any case, that doesn't matter. What does, is what it did to this country. That danged thing lowered the level of Great Slave Lake a good three feet and drained some of the shallow lakes around it. I can show any man who will pay attention where the water used to be right down below my cabin. The dam made a difference to the fishing and trapping out here. Also, some of the areas where ducks used to nest were drained. And to think this was done by intelligent people.

Another thing that bothers an old bull-headed man of independence like me I've mentioned before, and that's the existence of natives on relief. There's no reason in the Northwest Territories for any man in his prime to be on relief. The natives could get along fine in the same way their ancestors did before them. There are 500 miles of Barrens clear to the coast with no one in it. And fur is sky high. Right today, a man could get rich on fur. With prices like they were last year (spring, 1972), two men could go out to the Barrens and make $25,000 in one season. And they wouldn't have to work hard to do it because the fur is out there just for the taking.

Yes, there's fur dying of old age and the natives won't go after it. They would have it to themselves, too, because all the white trappers have come and gone, pulling out when white fox fur became cheap a few years ago. Most are now either too old or have passed on. It's a paradise out there right now.

Also, a newcomer had just as well forget it because he can't get a license to trap. I have one, but it doesn't do me any good because I'm too old. If a Canadian from outside the Territories came here wanting to trap, he would have to be a resident for six months before he could apply for the license, and then there'd be no guarantee that he would get it.

Should a man from the States come in wishing to trap, his chances would be pretty slim at best. First, he would have to become a Canadian citizen. Then, after residing in the Territories for six months, he could apply for a license. But, again there would be no assurance that he would get one.

One way I suppose a man could get to trap would be to marry a native woman. She could hunt and trap the year round if she so desired, and could take a white man along. As far as I know, this arrangement would be legal.

The government is tough about trapping licenses. If a year goes by and a trapper fails to buy his license, he has to get a certificate from a doctor explaining why he was unable to trap in that time. Otherwise, the license is lost and it's almost an impossibility to get back. One of the Peterson brothers suffered this fate when he left the country for several years. He had lived and trapped on the Barrens for a long time before that, but when he tried to get his license back, the government refused to grant it. I don't think that is completely fair, either.

CHAPTER XVI

Some of the happiest times during all my trapping years involved the preparations for another go at the Barrens. Good and proper equipment had to be gathered and put together in the fall with utmost care, because there was no turning around or coming back if something was broken, lost, or forgotten. Items and lists were checked and double-checked. This was fun, and I took pride in striking out with the best outfit I could find.

Good dogs were essential. I made my teams up with six or seven top-notch individuals. I purchased some of them and raised others from pups. A good dog was something a man never parted with. With the dogs secured, a good sleigh was next: an eight-footer and a cariole.

Two good rifles were needed, possibly a .270 or .308 with even the 30-30 a possibility. All shooting out in the Barrens, whether for caribou or wolves, was done right out in the open. Much of it was at long range, so a high-powered rifle was a necessity. The idea for having two rifles was plain enough: a man is out there where he can't get another if something goes wrong with one. More than one man has regretted not having a spare. And at least ten boxes of shells should be taken for each rifle.

Besides the big rifles, a .22 has its place for shooting ptarmigan. One could also kill live wolves in traps with it instead of using an ax, which wasn't very clean. Sometimes, a .410 shotgun was taken along for ptarmigan and ducks.

A good set of dog harnesses was also quite important—one that would hold up under the stress of cold weather and the strain of a heavy load. One set I had lasted me for ten winters. The good ones won't freeze in cold weather like the ordinary black harness that will break when it gets very cold. Dogs tend to bite this cheap kind, and patching it was a danged nuisance because breaks usually occurred right when a man least expected it.

A good snow shovel is critical. I maintained 150 traps—experience had proved this was all I could effectively take care of during the course of a season. For foxes, I used number ones; for wolves and wolverines, I used number threes and fours.

Before breaking camp in the spring, I always hung my traps up inside the cabin after boiling them if there was time. One winter, I even got the idea that the traps might work better if they were dipped in whitewash. This seemed to work pretty well, too. If the trap happened to become exposed after a storm, the whitewash was still on it, causing it to blend with the snow. I caught several foxes without even having to cover the trap. Still, covered traps worked the best.

Tents were just as important in the North as a house in the city is. I took two seven-by-seven-foot canvas tents with four-foot walls made of ten-ounce material that kept the wind out. These were usually made to order because the standard tents on the market in those days were inferior, with seven-ounce sides and eight-ounce tops. For Barrens use, that simply wasn't good enough.

Those sturdy little dwellings were used as out-camps though they weren't good enough by themselves. A frame was first built, then caribou hides were sewn together and stretched over it—with the fur out and the leather in. The tent was then pitched over the top of the hides and frame. The structure was put right out in the bald open, away from everything. If a man erected it by a ridge, he would come back after a bad storm and not find it. Snow would drift over the top, as I found out several times. Believe me, digging out a buried tent from beneath a snowdrift is not fun. But the tent would stand up and could take anything the Barrens had to offer.

When a man ran out of fuel at an out-camp, he had to lay in his bed to stay warm. Therefore, a good eiderdown and blankets needed to be taken along. Good hunting knives and a jack knife, boards for making fur stretchers, and a couple of seven-by-nine tarps for covering supplies in the fall when the plane lands were also staple items. Other standard items included a couple of good trail axes, a pair of snowshoes, spare rawhide and sinew called babiche for webbing, and a lot of one-fourth – and three-eighth-inch rope. The bigger rope was used to lash the top of the sleigh, the turn on the toboggan that's fastened to the lazy back and served as the ground lashing of the sleigh. Seventy-five to 100 feet of the quarter-inch rope was taken with its primary use being to serve as a main line running up to the lead dog.

The types of clothing a trapper wore ran a broad spectrum and was strictly a personal thing. But good caribou skin pants, and a parka, along with a pair of mukluks and mitts were standard. Heavy woolen underwear, shirts, and socks were my choice, and I avoided anything made of nylon.

You needed two fish nets. Nylon nets seemed to catch twice as many fish as the cotton variety; fish just couldn't get out of a nylon net. But you had to take care of your nets. In the fall when the fishing was done, we would take the nets from the lake and, after removing the floats and sinkers, thoroughly wash them. I never used factory-made lead sinkers, because they were forever getting tangled up in the net, making it useless. My net was sunk with fist-sized rocks, which caused it to hit bottom where it was meant to. Also, I made my own floats. After the net was prepared for storage, it was hung either inside or outside the camp, ready to be used again the following year. Perhaps a little mending was necessary, but that was pleasant work, and helped pass time early in the fall.

A canoe was needed for fishing the net. A vessel of at least fourteen feet was preferred. The one we had at Beaverhill Lake had been brought in by another trapper years before. It was in bad shape, but we managed to put a canvas on it and make it seaworthy enough to serve its purpose. We also had a small, home-made twelve-footer, so we were well equipped, as far as boats and canoes went.

All of this equipment would last for years if properly cared for. For example, some of the traps I was using during my last year on the Barrens had been in use for twenty years and were still in good shape. Once in a while, a trap was lost when a fox or wolf broke a chain or toggle and it had to be replaced. But it was unusual for a man to have to bring out an entire load of new traps.

FOOD LIST FOR TWO TRAPPERS, SEPTEMBER THROUGH MARCH

150 Lbs. White Flour

50 Lbs. Whole Wheat Flour

50 Lbs. Rice

100 Lbs. White Sugar

25 Lbs. Dehydrated Potatoes

50 Lbs. Dry Milk

50 Lbs. Brown Sugar

10 Gals. Gas for Main Cabin Light

Molasses by choice

Smoking Tobacco

Syrup by choice

1 Case Tea

1 Case Jam

1 Case Dry Fruit – Prunes

1 Case Dried Carrots,

1 Case Butter

1 Case Lard (for bannock)

2 Slabs Bacon

1 Case Coffee

1 Case Matches

Candles

Salt and Various Seasonings.

When I ran out of candles—the only light available at the out-camps—I'd make a bitch. This was just a flat dish or saucer into which I melted lard. I'd dip a wick or string into it and then light it. This gave off a nice light, and lasted for a long time. Seal oil was also good, and caribou fat would work in an emergency.

Whenever I was lucky enough to kill a bear in the fall, bacon was no problem. Thin strips of bear meat were just as good as any bacon on the market. With all these supplies and whatever nature provided, a man could live just as well on the Barrens in a trapping camp as anywhere else in the world. And he had a lot fewer worries and sickness to go along with it.

Most of the menus at the trapping camp were fit for a king, and had enough nourishment to give a man all the stamina he needed to stay warm and do his work. Fish were unequalled for flavor, and were fixed in a number of ways.

FISH CHOWDER

> Trout, cleaned and cut into chunks, and then put in a pot with enough water to cover it
>
> Throw in whole onion, or dried onions
>
> Pepper and salt to taste
>
> Add a pinch of sage
>
> Let simmer, adding butter and dried milk if available
>
> Boil only for a few minutes or the fish will crumble up and the whole thing will look like gravy
>
> Mix a little flour and water and just before the chowder is done, and add this to thicken

This is A-one food and hard to pass up.

Another way to fix a trout for a change resulted in one of my favorite meals. Not making any difference what size the fish was—two or three pounds or up to fifty pounds—it was cleaned and then put in the oven with just a little water over it. After it was thoroughly salted, it was roasted to a golden brown. When it was done, it was taken out of the oven and all the meat was removed from the bone with a fork, and put into a bowl. Some pre-boiled rice was mixed with the meat. Then flour, sage, pepper, and butter were added. Patties were made from this, and stored in a cool place until needed. They would keep for a long time, and when a man got hungry, some of them could be fried. Talk about a meal and a half! Instead of rice, potatoes, onions, or any combination thereof could be used in the patties.

Caribou meat was usually eaten as steaks, roasts, or in stew. But other ways of fixing the meat were possible with storing, preserving, and moving it around making it necessary to eliminate the bones. Pemmican and jerky were both excellent to the taste and easy to carry.

Cutting meat for jerky was done in a special way. The meat had very thin tissue separating the layers of muscle. From the natives who were quite efficient at this, I learned to cut the meat along this layer of tissue, never slicing through the muscle itself, but more in a lengthwise fashion. It was continually cut and rolled until, by the end of the quarter, a slab of thin meat a couple of feet long and several inches wide remained. This thin piece of meat held firmly together without flaking off. Learning to do this required a lot of patience and was an art.

Afterwards, the flat strips of meat were suspended from poles with a fire built below for smoking it. This could be done inside during the winter, with the poles placed near the stove. The meat, whichever the season, would turn black and dry with no danger of spoiling. After it had dried, it was pounded with a hammer or rock until it became as thin as possible, just short of the point where it would break apart. The next step called for putting it into a baking pan that was placed in a hot oven for about five minutes. When little pieces could be broken off, it was removed from the oven, and had become jerky. The meat was either hung up or stored in bales. Delphine specialized in making jerky which she kept and carried in a caribou skin bag. If she found the time, she would jerk four or five caribou and keep the entire amount in her bag.

Although pemmican was entirely different in the end result, the meat was cut and dried by the same process. Eventually pounded to a pulp like flour, Delphine combined it with the marrow from caribou or moose bones, occasionally adding such things as cranberries and currents for a special flavor. It was delicious, and could be eaten with a spoon. Three or four mouthfuls would provide enough energy to last a man a long time. It was also great as a quick snack while traveling, and I could live on the stuff. Over the years, pounded meat or pemmican was the Natives' long suit, keeping them going when there was nothing else.

The last few years I spent in the Barrens, I had a radio which kept me in touch with the world to the south. Even though I was usually tired and hit the hay early, that thing gave me a lot of satisfaction. Sometimes, our friends would talk to us over the air—CNRV, I think it was—and it was nice to hear from them. It was just like being back in civilization overnight. Many a night I slept in a doggone snow bank when trying to make it to the radio at the main camp for the Saturday night news from Edmonton or Winnipeg. The Northern Messenger was beamed to the North at 11 on Saturday night, and I always sure tried to hear it.

I have slept in the Arctic Barren Land several thousand nights and have to disagree with the findings of a group of scientists who went out from Churchill to study the Aurora Borealis. They reported that there was no sound associated with them, and I've certainly seen this same thing on other accounts. But on very cold nights, I have heard a noise like running water or a gentle wind rise up when the colors come close to the ground. There's a slight whooshing sound, and most of the trappers would confirm this. And they were a fantastic sight, as stream after stream spread across the sky, lighting up the earth across the Barrens. By the middle of August when the nights become noticeably longer, the northern lights begin to grace the heavens on moonless nights.

Recently, I read a book whose author stated that men of the cold Arctic never wore beards because of the moisture which froze in the whiskers. Although I didn't grow a beard, keeping mine rough-cut with a pair of clippers, many of the other trappers, my brother included, wore long, full beards and got along fine. I told them I'd grow hair where nature intended: on my head. I had one hair on my chest, and even though it was frequently pulled out, it kept growing back.

Another problem that a Barrens trapper faced was bathing, a task requiring some courage when the temperature had dipped to fifty or sixty below outside. But a man had to stand himself and a frequent bath did something for his morale. Here, the bath water was heated in a four-gallon square gas can kept for just this purpose. Getting clean was an accomplishment, but it made a man feel just as good there as anywhere else.

In talking about the Barren Land, trapping its expanse and hunting its animals, it would be incomplete not to mention the most unusual animal—the musk ox. Even though I never ran into them in great numbers in my travels or experiences, they were nevertheless out there, and I saw enough to learn a little about them.

When I was a youngster down in Manitoba, I had read and heard stories about this throwback to prehistoric times, and become fascinated with his nature and ability to live year-round in the harsh country. Unlike the caribou, he didn't migrate, living out on the tundra twelve months a year. I never dreamed that I would someday live and work in musk ox country.

If something lived or happened in the North, I generally managed to dig up information about it when I was a boy. Among that stash was a series of pictures of musk ox hides drying in the wind on the Arctic coast that I had found. It was soon after the turn of the century, and the season was closed on the animal, though the natives were permitted to take the meat and hides for their own use. At the time, I believe the hides were worth around $100 apiece.

Their range was along the Aylmer, Clinton-Colden, and Artillery Lakes, and out along the Thelon River. That was all musk ox sanctuary, but not all musk ox country. They seemed to hang around the mouth of the Hanbury River and Grassy Island (about ten miles below the mouth of the Hanbury) in the summer.

Occasionally, an odd herd strays away from the Hanbury, and it was one of these that Phil and I happened to see. There were twenty-two of them, and not far from our camp. Not hanging around very long after spotting us, they were like a bunch of cattle. We spotted the odd bull a while later, but other than that, I saw very few.

A few years ago, a trapper discovered a herd of twenty-five musk ox wintering on the shores of Clinton-Colden Lake and, three years ago, the pilot of a water resources plane spotted three at the south end of Artillery Lake, where the game warden was also aware of them. This was nearly 100 miles south of their range. They'd apparently followed the shore of the lake until they nearly reached the tree line.

In a herd, the bulls stand on the outside with the cows and calves protected in the center. You would think the wolves would have trouble getting to them, but a portion of the herd falls to this predator. A biologist I knew spent a summer studying the musk ox, and found a big bull that had been killed by wolves on the banks of the Thelon River. There had been quite a struggle, and after studying the bones to determine the bull's age, he said there was no doubt about the wolves having killed him. Wolves were killed by the biologist for studying their stomachs, and he found small bones from the backbone of musk ox inside these wolves. Calves were the most frequently killed, with wolves slipping in to do their business, even in spite of the cows' efforts to cache their young.

Once, a fellow from the Department in Ottawa stopped in Fort Reliance on his way south after being out on a musk ox survey to report that he had counted 1,400 head in the area he'd been surveying. Their numbers seem to be on the increase, and there's a rumor that a dedicated season is going to be opened for trophy hunters. I am strictly against this because this beautiful animal is already hanging on by the skin of its teeth, and its extermination would be the biggest crime to ever hit the North. I wouldn't shoot one if it jumped over the top of me.

Once a man has trapped the Barrens, the feeling for it never leaves. There is a pull and a mysticism about the place that never lets a man out of its grasp. Not long ago, I was talking with an old friend, a game management officer who once trapped for a living. He had just returned from a caribou survey and said that he'd seen one of the largest concentrations of caribou in recent years. They were around Great Bear Lake, headed southeast toward Fort Reliance. This brought back memories of the big herds in September out beyond Whitefish Lake and out into the Thelon. When I asked this man what he was going to do when he retired from game management, he said he was going back to the trap line. He echoed my thoughts that a man could make a fortune with present-day prices.

Blizzards were never a big worry for a trapper, because a man got used to them and came to accept them as part of the make-up of the country—a natural hazard that went along with the game. Mind you, they could get tough if a man became careless and let panic take control of his being. A native friend who got caught just a couple hundred yards from his camp had once gone to a lake nearby for a pail of water and, by the time he'd cut the ice hole and started back, a storm had hit. Unable to see his camp through the flying fury, he'd squatted in the snow rather than getting lost. He was dressed in warm clothing, and had to stay there all night. He smoked cigarettes—and, believe me he went through a lot of them that night. The next morning, it had cleared and he made it back to camp, pretty well chilled, but alive, and all because he hadn't panicked.

A similar thing happened to me one time when a neighbor on the Barrens asked me to spend a night with him. I had tried to find his place, but couldn't, and so had turned around to avoid camping out. I saw him again a few days later and he issued the invitation again. The results were the same, but this time I got caught in a dirty storm and had to stop. There was dog food and I had a tarp to use for a bed, but I was afraid to sleep for fear of freezing to death. The dogs were hitched so I'd be ready to travel at first light. When it arrived, I'd looked up and his bloody camp was only a quarter mile away. It sure made me feel foolish. Another thing I learned on that trip was never to trust a man's judge of distance. When talking about a place, one man might say it is ten miles away and another fifteen. Trapper miles varied with the individual and, in most cases, you simply didn't know how the hell far away a place was.

Two memorable hunts come to mind, with one of my brothers involved in each. They were terrific shots and outstanding hunters who had the kind of thirst for the sport that keeps men optimistic. The one with Hughie came when we were paddling downstream on the way to our proposed camp below Granite Falls on the Thelon. We'd spotted a herd of seventeen caribou on a small island in a lake. Maneuvering to the downwind side, we'd landed, with Hughie staying behind and crawling toward a rough hill as the caribou had gone to the other side of the island. I'd circled back to the herd and managed to get one shot as the herd trotted back toward Hughie. With his .250 equipped with a four-power scope, he'd opened up on them. Unable to see him, sixteen of them dropped before the shooting ceased. All this time, I'd been concerned, because he wasn't sure where I was and some of the bullets had struck the gravel just a few feet from where I lay flat on my face. Caribou were funny that way. When they couldn't see the rifleman, they trotted back and forth, unsure of which direction to flee.

We cached the meat on this little island by hanging it on tri-poles and letting the wind and sun to dry it to form a thick crust on the outside. It kept well and was good for eating meat. This was at the end of my three-day line, so I had plenty of meat there to last me. Killing those caribou was legal, and still is to the holder of a general hunting and fishing license.

The other hunt came while I was trapping north of there with Phil at a time when we really needed meat desperately. It was September 20, and the big herds were still out on the tundra to the north, their summer range. So we were out on this cold, frosty morning when we'd spotted two bulls and a cow about a quarter mile to the east with a small herd lying down a little beyond. The country was very flat, and we'd walked right toward them. Pretty soon, the three had come toward me and, once in range, I'd dropped them immediately, taking the insides out and caching the lot. When I'd started shooting, the resting herd had taken off, accompanied thereafter by shooting from Phil's end. From where I was, I could hear the bullets making that dull thud as they connected with their intended targets. Phil had scored; in fact downing the whole herd, which numbered an even dozen. Between us, we'd bagged fifteen in just a few minutes.

The big herds had come through soon after, and we'd gotten all we needed for the winter. One morning while doing so, we spotted four bulls about a mile away, identifying them by their white manes because the distance was too great for us to make out their antlers. With no wind, we stood still as they came onward. Spotting us, they turned abreast. I chose the two on the left, leaving Phil the others. These beautiful animals were in their best with tremendous racks on their heads. In four seconds and with four good shots, the bulls were ours. To the hunter-trapper, there was no better sight on a cold, frosty morning.

The years slipped by as the cold winds came and covered the tundra with driven snow, the tracks and marks I'd made beneath them spanning thirty years. Although my life in the Barren Land was brief compared to the ages, it marks the end of a golden era of freedom-loving men who pursued fur and game just for the sake of the treasure and adventure of it all. Names such as Murphy, Nelson, Stewart, Croft, Grodsky, Duhamblond, Magrum, Cooley, Peterson, Price, MacKay, Pearson, Langstaff, Knox, Greathouse, Stark, DeSteffany, Riddle, and of course D'Aoust were found yearly spread across the great expanse of the Arctic tundra. This has all ended with who knows what to replace it. Not only are those men absent from the lonely, wind-blown trapping camps, but trapping as we knew it no longer graces this great country.

Toward the end of my trapping days, and after a rough return trip to Fort Reliance, the Mounties tried to talk me into quitting the trap lines. One man in particular worked to convince me to retire. He asked me why I didn't hang it up and call it a day. In a friendly way, he told me to put the traps away because I was too old and things were getting too rough out there for me.

I agreed with him that it was getting tougher each year, but said I figured I hadn't slipped as much as he thought. I told him that I would think about what he said, then crossed the bay to my cabin with Delphine. We immediately started making plans for the next year, and these plans all centered around returning to the Barrens. I was over sixty.

When the Mountie heard about this, he just shook his head and said that I'd probably be too bull-headed to ever quit. I couldn't hang my traps up because there was a living to be made, and just what the hell could I do besides trap? I had tried working at a job, but was no good for that kind of life.

I didn't stop trapping either, and made a couple of good stakes after that. When the Mountie moved to Vancouver, he wrote to me to ask what in the world made me tick.

In the end, he was right, and I did retire from my chosen trade just a few years later—about a dozen years ago.

The human mind is a wonderful thing, as a man can look back and recall events all the way to his childhood. Even though I'm now a thing of the past who has just recently started to think of the end, I'm still free and happy, with plans for the future and thousands of fond memories to keep me going. For more than fifty years—thirty in the Barrens—I have spent my life hunting and trapping, and have always been drunk with the excitement, never sober. Now that the years are nearly gone, I can look back to a life I loved. The magnetic North still claims my soul.

There's a little spot at Fort Reliance where it is quiet, but not lonesome. That one spot—just about four miles south where the caribou portage into a lagoon and then into McLeod Bay—attracts me particularly. Thousands of caribou have made that portage, headed north to the Barrens to summer. This spot is on Maufelly Point, a peninsula jutting north and toward the narrows a half mile wide which finds Fairchild Point sticking out to the south and on into North Bay. This is rough, rolling land, marked as typical of Great Slave's east end. The portage has plenty of small and medium spruce. The aircraft fly over it going to and from Fort Reliance and Yellowknife. Maybe the clattering of caribou hooves and the hum of the aircraft from above will always be a reminder.

I mention Maufelly Point, a four-mile peninsula opposite Caribou Islands, four miles from Fort Reliance. This picturesque spot that I love is sandy, a beautiful area overlooking Charleton and McLeod Bays. If my friends want to take the trouble when the time comes, I'd like to be left there.

And if I were given the right to start all over again, I'd be back on the job. I went in for the adventure side of it, although there was a good living all those years, and was never disappointed. Father Time is rolling along, and I'm content to sit back now and look at memories and experiences I had out there. There were thousands of campfires kindled on those trips by scow, canoe, boats, dog teams, and, in later years, plane. I cherish all of this, and look back on it as a life spent doing something I wanted to, always as a free spirit.

✻

*

CPSIA information can be obtained
at www.ICGtesting.com
Printed in the USA
FFOW02n0219050416
22921FF